Designers' Guide to EN 1992-1-1:2023: Eurocode 2

emerald PUBLISHING

ice Publishing

Designers' Guide to EN 1992-1-1:2023: Eurocode 2

Strengthening of Existing Concrete Structures with Carbon Fibre Reinforced Polymer (CFRP)

Kaloyana Kostova and Craig Giaccio

Published by Emerald Publishing Limited, Floor 5,
Northspring, 21–23 Wellington Street, Leeds LS1 4DL.

ICE Publishing is an imprint of Emerald Publishing Limited

Other ICE Publishing titles:
Concise Guide to Reinforced Concrete Design to Eurocode 2
Patrick Purcell. ISBN 978-0-72776-572-7

ICE Handbook of Concrete Durability
Edited by Marious Soutsos. ISBN 978-0-72776-375-4

Fibre-reinforced Concretes for High-performance Structures
Andreas Lampropoulos. ISBN 978-0-72776-556-7

A catalogue record for this book is available from the British Library

ISBN 978-1-83549-935-1

Cover photo: Maksim Ladouski/Alamy Stock Photo

Commissioning Editor: Michael Fenton
Content Development Editor: Ryan Molyneux

Books Production Lead: Benn Linfield
Typeset by KnowledgeWorks Global Limited
Index created by Lyn Nesbitt-Smith

Printed and bound by CPI Group (UK) Ltd, Croydon, CR0 4YY

Contents

Preface

Scope and field of application

The information contained in the informative Annex J of *BS EN 1992-1-1:2023* (BSI, 2023b) is intended to provide the designer with rules specific to the strengthening of structures with adhesively bonded carbon fibre reinforced polymer (CFRP) strengthening systems.

The standard covers the following adhesively bonded reinforcement (ABR).

- Externally bonded reinforcement (EBR) systems, including
  - prefabricated carbon fibre reinforced polymer (CFRP) strips
  - carbon fibre (CF) sheets or pre-impregnated laminates.
- Near surface mounted (NSM) systems comprising either
  - NSM CFRP strips
  - NSM CF bars
  - NSM CF rods.

In the vast majority of cases, the design rules for CFRP link to the provisions in *BS EN 1992-1-1:2023* as a whole to ensure consistency. Instances where this does not occur are acknowledged in this book, and an explanation provided for why this has occurred.

Strengthening of concrete structures with CFRP has become fairly common around the world. However, limitations on available testing results to cover all scenarios encountered in design became evident in the development of design rules contained in Annex J.

No provisions were given in the second generation Eurocode related to design of CFRP strengthening systems for fire resistance.

Aim of this guide

The aim of this book is to provide practising design engineers and other code users with examples of the implementation of the provisions of *BS EN 1992-1-1:2023* pertaining to strengthening of concrete members with CFRP. It also aims to provide an understanding of some of the basic principles used in *BS EN 1992-1-1:2023* regarding strengthening of structures with CFRP systems.

The book is aimed at practising engineers who have a sound grasp of structural engineering principles, in particular the principles that apply to reinforced and prestressed concrete design. It does not set out to reproduce significant volumes of information pertaining to manufacturing methods of CFRP systems. It does, however, include pertinent information where required to support the basic understanding as to how the system works or why certain approaches were taken in the development of the standard, as well as key points about the system and its execution that may influence design decisions.

In writing this book, the authors observed that some additional content could be provided to help designers implement the provisions in *BS EN 1992-1-1:2023*. They have sought to provide this additional content, in particular noting the source of where this has come from. This is intended to inform readers where additional content would support undertaking design, and provide the authors' opinion of how they would resolve any gaps. Designers and other code users are encouraged to make their own assessment of recommendations made in this book outside of the provisions of the standard, and to make their own informed decisions as to how to implement the provisions of the standard.

Layout of this guide

The British Standard *BS EN 1992-1-1:2023* has a national foreword and 14 sections, together with 19 annexes. Annex J, which is the subject of this design guide, consists of 15 subsections numbered J.1 to J.15.

The layout of the guide is not aligned with the structure of *BS EN 1992-1-1:2023*. Instead, it uses Harvard referencing style and italics font to cite relevant parts of the code. Quotations from other sources, including other Eurocodes, and cross-references to sections, etc., of this guide, are in roman type. As the guide contains five annexes, it must be noted that none of them is aligned with a particular annex in *BS EN 1992-1-1:2023*.

When the standard is first introduced in a chapter, full in-text citation is used, i.e. *BS EN 1992-1-1:2023* (BSI, 2023b). When repeated, in-text citations are shortened to *BS EN 1992-1-1:2023*. Sections, clauses, expressions, figures and tables usually follow the code citation separated by a comma – for example

BS EN 1992-1-1:2023, J.5;

BS EN 1992-1-1:2023, J.11.1.2.5;

BS EN 1992-1-1:2023, Formulae 8.56 to 8.62;

BS EN 1992-1-1:2023, Figure J.3;

BS EN 1992-1-1:2023, Table 4.3.

Alternatively, the cross-reference may be described as part of the code – for example, Section 3 of *BS EN 1992-1-1:2023*.

Acknowledgements

The authors would like to acknowledge the contribution of Professor Antony Darby from the University of Bath, UK. Professor Darby has provided valuable insight into the background on a number of areas in this book and supported the authors by reviewing this material. His contribution has been very valuable.

The authors would like to acknowledge and thank Concrete Repairs Limited for providing a selection of photos for use in this book. This has enabled the authors to provide what, in their view, is a well-balanced combination of theoretical descriptions and real photographic descriptions of the material, applications and unique considerations that apply for strengthening concrete structures with CFRP.

Craig would like to extend his gratitude to his family for their unwavering support, patience and encouragement throughout the development of this work. Without their understanding during long hours of research and writing, completing this book would not have been possible. Craig's wife's constant belief in his work and steadfast support has been a source of strength. His daughter's curiosity and enthusiasm has reminded him of the importance of perseverance and advancing knowledge.

About the Authors

Dr Kaloyana Kostova PhD CEng FICE is a Fellow of the Institution of Civil Engineers with practical and research experience in the field of structural engineering. She holds a PhD in Civil Engineering from the University of Bath where her studies were focused on the use of fibre-reinforced polymer (FRP) reinforcement in structural concrete cast in fabric formwork.

In 2017 Kaloyana joined the Eurocode development project team working on the addition of new items in the next generation of EN 1992-1-1, including FRP strengthening of reinforced concrete structures, use of FRP embedded reinforcement, stainless steel reinforcement, assessment of existing structures and steel fibre reinforced concrete (SFRC). She continued her involvement in the development of standards as a co-opted member of the B/525/2 committee on 'Structural use of concrete', a member of the ISO/TC 071/SC 06 committee on 'Non-traditional reinforcing materials for concrete' and a member of fib Task Group 5.1 on 'FRP reinforcement for concrete structures'. She also has a strong interest in research on fire safety and durability of FRP composites as key questions in the adoption of these materials.

In 2021 Kaloyana joined the National Composites Centre (NCC) in the UK, where she leads a structural analysis team delivering innovative FRP composites design for different sectors, including construction, wind, hydrogen, aerospace and marine. Prior to her deep dive in exploitation of composites at NCC, Kaloyana spent most of her career as a bridge designer working for Tony Gee and Partners and, earlier, for Scott Wilson. Her experience varied from bridge investigations, assessments and strengthening schemes to new bridge designs, including FRP strengthening of existing structures and new FRP composite bridges.

Kaloyana first explored FRP composites for their potential as a reinforcing material for concrete during her PhD studies under a Leverhulme Trust grant on 'A new form of architecture for concrete structures'. She developed an optimisation tool for structural analysis of concrete formed using an innovative flexible formwork system and reinforced with various types of FRP due to their flexibility and light weight, as well as offering a novel end anchorage for FRP bars. She also had an early experience in design and construction of buildings, following her graduation from the University of Architecture, Civil Engineering and Geodesy in Bulgaria.

Kaloyana is actively involved in developing capabilities within industry, dissemination of knowledge and improving regulations, as well as creating methods and technologies aimed at shaping the future of engineering.

Dr Craig Giaccio BE Civil (Hons), PhD (Eng), FICE, FIEAust, CPEng, NER, APEC Engineer, IntPE(Aus) is a chartered civil engineer with over 27 years of experience that spans design delivery of major infrastructure projects through to development of codes and standards for structural engineering. Craig is a Fellow of the Institution of Civil Engineers and the Institution of Engineers Australia. He is presently a Technical Director at Hewson Consulting, UK.

He is currently Deputy Chair of the B525/2 committee and the National Annex Group responsible for preparing the National Annex and Published Document for *BS EN 1992-1-1:2023*. This commission follows a previous commission where he led a project team developing material for the new concrete design standard. He is also an active member of fib Task Group T5.1 on fibre reinforced polymer reinforcement.

Craig has played a key role in the delivery of an array of structural and civil engineering projects. This includes major infrastructure projects such as Queensferry Bridge and HS2. He has additionally undertaken a significant number of commissions involved in assessment and remediation of aging concrete structures in the UK and Australia.

Additionally, Craig has a track record of supporting the development of the profession. He was the Chief Editor of the *ICE Bridge Engineering Journal* from 2017 to 2023, and reviews technical submissions for the *ACI Structural Journal* and the *Australian Journal of Structural Engineering*.

List of Symbols

Where possible, symbols have been adopted from BS EN 1992-1-1:2023 (BSI, 2023b). In some instances, additional symbols have been used for descriptive purposes as outlined below.

A_{equiv}	Equivalent effective area of tension steel of a member strengthened with CFRP with the CFRP transformed into an equivalent steel layer and all tension reinforcement layers combined as a single layer
$A_{f,eff}$	Transformed CFRP area to steel
$A_{f,1}$	Area of a single CFRP sheet
$A_{sc.i}$	Area of compression steel of the ith layer of tension reinforcement with $i = 1$ corresponding to the layer of compression reinforcement with the highest compression stress for members in flexure
$A_{st.i}$	Area of tension steel of the ith layer of tension reinforcement with $i = 1$ corresponding to the layer of tension reinforcement with the highest tension stress for members in flexure
F_c	Compression force in concrete
F_{ct}	Tension force in concrete
F_{equiv}	Combined tension components of the cross section
F_f, F_{fEd}	Force in the CFRP reinforcement
$F_{fEd,max}$	Force in the CFRP reinforcement generated by the lower value of fatigue loading
$F_{fEd,min}$	Force in the CFRP reinforcement generated by the upper value of fatigue loading
F_s	Force in the steel reinforcement
F_{st}	Design force in the CFRP shear stirrup resisting end cover separation
$\Delta F_{fEd,max}$	Maximum value of $\Delta F_{fEd,fat}$ between two cracks under the fatigue load combination being considered

$\Delta F_{fk,B}$	Bond resistance between cracks
$I_{cr,post}$	Post-strengthening cracked second moment of area (moment of inertia)
$I_{cr,pre}$	Pre-strengthening cracked second moment of area (moment of inertia)
$M_{Ed,shift}$	Design value of the applied bending moment after applying the shift rule
$M_{Ed,SLS}$	Design value of the applied bending moment at SLS
$M_{Ed,str}$	Design value of the applied bending moment at strengthening
M_p	Prestressing bending moment
M_{post}	Additional applied bending moment after strengthening (equal to the difference of the total applied moment and the moment applied at strengthening)
$M_{Rd,pre}$	Design value of bending moment resistance of the unstrengthened reinforced or prestressed concrete member
ΔM_{Ed}	Applied bending moment difference between cracks
$N_{Rd,pre}$	Design value of axial resistance of the unstrengthened reinforced or prestressed concrete member
T_g	Glass transition temperature
$a_{\delta,post}$	Total deflection generated by the additional applied bending moment after strengthening M_{post}
$a_{I,post}, a_{II,post}$	Value of $a_{\delta,post}$ calculated for uncracked and fully cracked conditions, respectively
$a_{\delta,pre}$	Total deflection generated by the applied bending moment at strengthening $M_{Ed,str}$
$a_{I,pre}, a_{II,pre}$	Value of $a_{\delta,pre}$ calculated for uncracked and fully cracked conditions, respectively

d_{equiv}	Equivalent effective depth of a member strengthened with CFRP strengthening with the CFRP transformed into an equivalent steel layer and all tension reinforcement layers combined as a single layer
d_f	Effective depth of CFRP reinforcement
d_{fv}	Effective depth of CFRP strengthening, measured from the top of CFRP shear strengthening to the steel tension reinforcement
$d_{sc.i}$	Effective depth of compression steel of the ith layer of tension reinforcement with $i=1$ corresponding to the layer of compression reinforcement with the highest compression stress for members in flexure
$d_{st.i}$	Effective depth of compression steel of the ith layer of tension reinforcement with $i=1$ corresponding to the layer of tension reinforcement with the highest tension stress for members in flexure
d_{st}	Effective depth of total steel reinforcement in tension
$f_{bfk,max}$	Characteristic maximum resistance of an end anchorage
f_{bfRd}^{*}	Adhesive bond resistance component of the intermediate crack resistance to be used for simplified fatigue design method
Δf_f	Stress change in CFRP under the fatigue load combination
$\Delta f_{fRd,\,fat2}$	Fatigue stress in the CFRP system subject to N* stress cycles
h_f	Height of shear strengthening
h_{max}	Maximum distance from centre of gravity to extreme concrete fibre
k_1	Concrete strength reduction factor
k_2	Factor for the depth of resultant compression force
n_s	Factor depending on the configuration of the CFRP shear strengthening system (fully wrapped or open)

n	Total number of CFRP stirrups crossing a crack
m	Number of CFRP stirrups crossing a crack with anchorage length less than characteristic maximum value of effective bond length
p_f	Bonded perimeter of the EBR CFRP system
$q_{Ed,SLS}$	Applied UDL at SLS
$q_{Ed,str}$	Applied UDL at strengthening
s	Slip of the CFRP reinforcement or spacing of steel shear reinforcement
s_{f1k}	Slip of the CFRP reinforcement at maximum bond stress
$s_{f,max}$	Maximum spacing of individual CFRP strips
z_m	Lever arm between tension and compression forces in a member subject to flexure for use with the simplified method for determining internal forces and stresses for intermediate crack debonding checks
a_{fE}	Distance between the end of the CFRP strengthening and the nearest point of zero moment (including a point of contraflexure in a hogging zone)
β	Angle of inclination of a CFRP shear strengthening relative to an axis perpendicular to the span of the member being strengthened
ε_{bot}	Strain in concrete at the outermost tension fibre
$\varepsilon_{cu,c}$	Ultimate compression strain of concrete confined with CFRP, which has value of 0.006 unless more accurate information is available
$\varepsilon_{c2,c}$	Compressive strain concrete, at which the bilinear stress strain curve for concrete confined with CFRP changes gradient, taken as 0.00175
$\varepsilon_{c,post}$	Strain in concrete generated by the additional applied bending moment after strengthening M_{post}

$\varepsilon_{c,pre}$	Strain in concrete generated by the applied bending moment at strengthening $M_{Ed,str}$
ε_{equiv}	Equivalent strain in tension reinforcement of a member strengthened with CFRP, with the CFRP transformed into an equivalent steel layer and all tension reinforcement layers combined as a single layer
ε_f	Strain in CFRP reinforcement
$\varepsilon_{h,rup}$	Hoop rupture strain of CFRP confining reinforcement
ε_{top}	Strain in concrete at the outermost compression fibre
$\varepsilon_{sc.i}$	Strain in the ith layer of compression reinforcement with $i = 1$ corresponding to the layer of compression reinforcement with the highest compression stress for members in flexure
$\varepsilon_{st.i}$	Strain in the ith layer of tension reinforcement with $i = 1$ corresponding to the layer of tension reinforcement with the highest tension stress for members in flexure
ε_0	Strain at the soffit of a reinforced or prestressed concrete member at the time of strengthening (also called CFRP pre-strain)
κ_h	Factor determining the contribution of clamping force to the intermediate crack debonding resistance
λ_{crit}	Limiting slenderness for isolated members below which second order effects may be neglected for CFRP strengthened columns
σ_{equiv}	Equivalent stress in tension reinforcement of a member strengthened with CFRP, with the CFRP transformed into an equivalent steel layer and all tension reinforcement layers combined as a single layer
σ_f	Stress in the CFRP reinforcement
$\sigma_{c,post}$	Stress in concrete generated by the additional applied bending moment after strengthening M_{post}

$\sigma_{c,pre}$	Stress in concrete generated by the applied bending moment at strengthening $M_{Ed,str}$
$\sigma_{s,post}$	Stress in steel generated by the additional applied bending moment after strengthening M_{post}
$\sigma_{s,pre}$	Stress in steel generated by the applied bending moment at strengthening $M_{Ed,str}$
$\sigma_{sc.i}$	Strain in the ith layer of compression reinforcement with $i=1$ corresponding to the layer of compression reinforcement with the highest compression stress for members in flexure
$\sigma_{st.i}$	Strain in the ith layer of tension reinforcement with $i=1$ corresponding to the layer of tension reinforcement with the highest tension stress for members in flexure
$\tau_b(s)$	Bond-slip law describing the relationship between the shear stress between the CFRP and concrete substrate and the slip required to generate this shear stress
x_{post}	Depth of the neutral axis corresponding to the additional applied bending moment after strengthening M_{post}
x_{pre}	Depth of the neutral axis corresponding to the applied bending moment at strengthening $M_{Ed,str}$

List of Figures

Kaloyana Kostova and Craig Giaccio
ISBN 978-1-83549-935-1
https://doi.org/10.1108/978-1-83549-934-420251001

Chapter 1
Introduction to fibre reinforced polymer strengthening systems for concrete structures

General

Fibre reinforced polymer (FRP) strengthening of concrete structures has become more commonly used over the last couple of decades. FRP strengthening systems rely on high-strength tension fibres contained within a fibre/polymer matrix to provide additional tensile strength to concrete structural members subjected to flexure, shear and torsion, or to confine concrete columns to increase compression resistance. Commonly used FRP strengthening systems comprise either carbon, aramid, glass or basalt fibres. Efficient use of FRP material is achieved by aligning fibres with the tension forces that are resisted in a structure being strengthened or those that arise from confining effects.

FRP strengthening systems obtain their structural properties from the fibres that are embedded within a polymer matrix. The surrounding polymer matrix binds the fibres together and facilitates load share between them. The surrounding polymer material can also provide resilience against accidental impacts as well as protection from environmental conditions such as ultraviolet (UV) light. There are a variety of systems on the market, some have uni-directional fibres and some have multidirectional fibres.

FRP strengthening is commonly achieved by adhering an FRP fabric or plate to the exterior surface or notching the surface to embed a bar or rod, as shown in Figure 1.1. Other systems that are used with some frequency include embedding FRP bars or rods within concrete members or fixing prestressed FRP to external surfaces of a concrete member fastened to a concrete surface with mechanical anchors. An illustration of the deep embedment technique is shown in Figure 1.2. The systems shown in Figure 1.1 are the only systems covered by *BS EN 1992-1-1:2023* (BSI, 2023b).

FRP strengthening continues to become more popular with time because it has a higher tensile strength than steel. This means that the weight of structural components used for strengthening is lower than it would be with steel, bringing installation benefits and adding relatively little dead load to a structure. FRP is also a very light and easy-to-transport solution compared to solutions such as steel plate strengthening.

It has disbenefits on cost. Material costs are generally higher than steel. Labour costs of surface preparation are additional to this. Therefore, the selection of the system needs to carefully consider all aspects of installation prior to ensuring its economic viability.

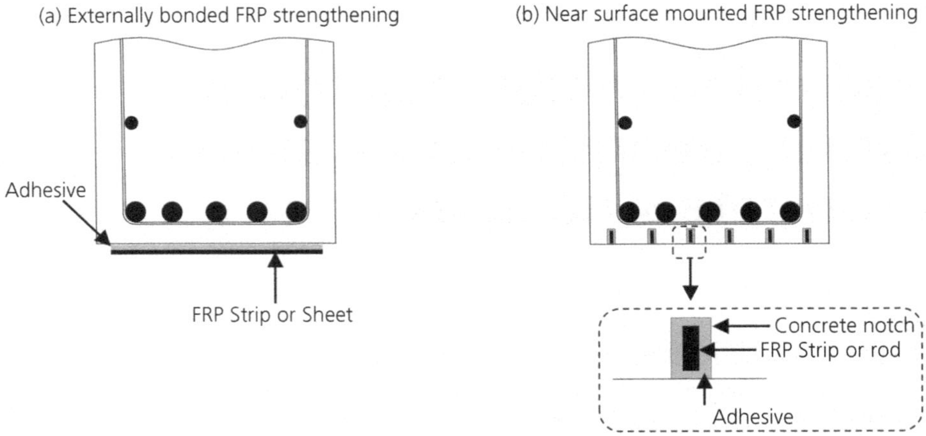

Figure 1.1 Illustration of beams strengthened in flexure with strengthening systems covered in Annex J (Adapted from *BS EN 1992-1-1:2023*)

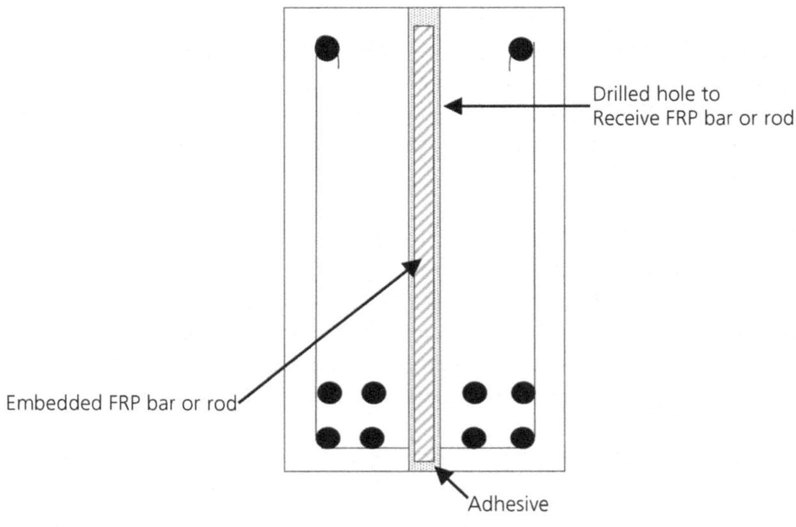

Figure 1.2 Illustration of shear strengthening with deep embedded FRP strengthening method

Standardisation in *BS EN 1992-1-1:2023* was undertaken only for carbon fibre reinforced polymer (CFRP) strengthening. This was simply because this application was the only FRP strengthening system that had a large enough database of necessary information to cover all aspects addressed in the code. Other strengthening systems commonly available on the market are glass fibre or aramid fibre reinforced polymer (GFRP and AFRP) strengthening systems.

Figure 1.3 CFRP strengthening systems covered by *BS EN 1992-1-1:2023*

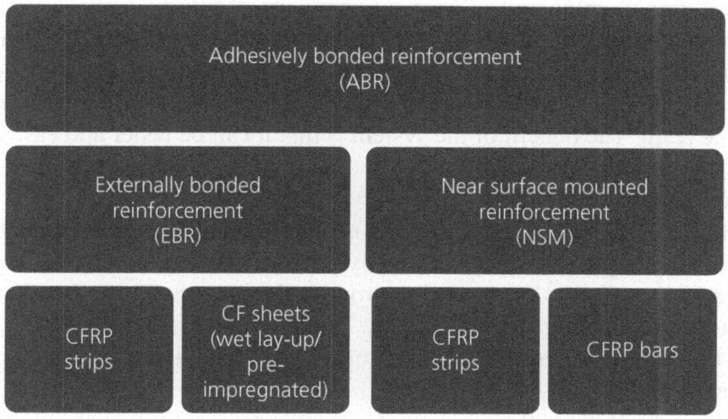

Strengthening applications covered by the standard

All CFRP strengthening systems covered in *BS EN 1992-1-1:2023* are adhesively bonded reinforcing (ABR) systems which are not prestressed. This standard covers the following ABR systems.

- Externally bonded reinforcement (EBR) systems which include CFRP bars, sheets and strips, bonded to the external surface of concrete with adhesive that provides bond to the surface as illustrated in Figure 1.1a.
- Near surface mounted (NSM) systems which are strengthening systems that require a slot to be cut into the concrete cover of the strengthened structure, into which a bar or strip is bonded as illustrated in Figure 1.1b.

Figure 1.3 outlines the terminology used in *BS EN 1992-1-1:2023* for CFRP strengthening.

Production methods

Al-Mahaidi and Kalfat (2018) explain that carbon fibres are created from a precursor material which is used to create filament yarns. The precursor material is commonly polyacrylonitrile, with also rayon or petroleum pitch being used. The filaments are taken through a heating and oxidisation process to create the fibres which are then coated to protect them from damage. These fibres are then woven into fabrics or pultruded into laminates (strips) or bars.

There are various production and installation methods available to produce a strengthening system from the bare fibres. Production of the CFRP strengthening material from the bare fibres varies. The systems covered by BS EN 1992-1-1:2023 (BSI, 2023b) are listed below along with a brief summary of how they are produced.

- CFRP EBR sheets.
  - Wet lay-up sheets – this is a system where a dry fibre matrix (either uni- or multidirectional) is impregnated with compatible saturating resin on site. The resin, along with compatible primers, adhere the CFRP matrix to the strengthened surface of the structure.

  This is a popular method of applying CFRP, owing to its flexibility, and can be used in a wide variety of applications, including systems that require wrapping around corners (e.g. shear strengthening and column confinement).
  - Pre-impregnated sheets – in this system, a dry fibre matrix is pre-impregnated with resin prior to transport to site. The fibre matrix is partially cured ahead of installation. On site, additional resin may be added when the sheets are installed, and heating may be required to cure the resin. The benefit of the system is that it can be faster and easier to install and provide shorter curing times.
- CFRP EBR or NSM strips.
  - EBR strips are pre-cured offsite, with the resin to bind fibres together being applied in a controlled factory environment. These systems can be formed to a variety of shapes. The pultrusion method is often used to produce these strips. The pultrusion process undertakes the following steps
    - The fibres are first fed through a preforming guide to achieve the required fibre orientation.
    - The reinforced fibres then undergo resin impregnation in a 'wet-out' bath of polymer.
    - The reinforced fibres with the resin are then passed through a heated die. In this process the fibres begin to solidify.
    - Near the end of the heating process the resin temperature will become higher than the die, and a separation process begins where the cured profile can be removed from the die, leaving the strips.
- CFRP NSM bars or rods.
  - Round bars are often manufactured using the pultrusion process described above.
  - Bars and plates are also manufactured using the pultrusion process described above.

Product specification

At the time of producing this standard, there was no harmonised EN standard to cover the production of the CFRP strengthening material. However, European Assessment Documents (EADs) were being prepared for the European Organisation for Technical Approvals (EOTA), which will enable European Technical Assessments (ETAs) to be produced. These will provide a basis on which suppliers can obtain certification of their product under the European Construction Directive (CE marking).

The intention of the drafting group who produced Annex J was to ensure that it would enable use of these ETAs when they become available. This was achieved by including statements referring to 'production data' or 'more accurate information' in the annex.

Additionally, these ETAs are likely to cover systems not included in the standard, such as deep embedded bars for shear strengthening, or externally prestressed systems.

At the time of writing this book, EAD 160086-00-0301 – Kits for the strengthening of concrete elements by externally bonded CFRP strips (EOTA, 2019) was not yet granted and, therefore, was not available.

ETAs provide manufacturers with a route to producing a CE mark for a product that is not covered by a harmonised European standard. It is the European Union (EU) Construction Products Regulations (CPR) that enables this route. A product that complies with an ETA can, therefore, be used throughout the European Union and any other country that utilises the EU CPR. Use of these documents in a country outside of the jurisdiction of the EU CPR may need confirmation of its validity for use or agreement to its use as part of a project specification.

Assessment for strengthening

Assessment of a structure for strengthening may be undertaken for a variety of reasons. As stated in TR55 (Concrete Society, 2013), this can include a change in load regime (e.g. change of use for a building or changes in bridge loading standards), to cater for changes in standards since a structure was constructed, to add reinforcement to an under-designed structure or one not constructed correctly, to address concerns regarding observed deterioration, or to add reinforcement for alterations (e.g. around holes cut into floor slabs in buildings to enable services or stairs to be added).

BS EN 1992-1-1:2023 includes a new Annex I dedicated to the assessment of concrete structures. The annex includes the following items

- methods to enable determination of characteristic material properties for assessment to avoid conservative properties being used
- adjusting partial material factors where testing is used to determine material properties
- deterioration
- calculating bond and anchorage lengths of plain bars in a way that is consistent with the method for ribbed bars
- dealing with detailing that doesn't conform to the current standard
- more rigorous assessment provisions for assessment of shear resistances of members without shear reinforcing using Critical Shear Crack Theory (Muttoni and Fernandez Ruiz, 2008)
- using the beneficial effect of compression concrete flanges to increase shear strength, shown in various sources–as illustrated in Giaccio (2003).

The methods in this annex should be exhausted prior to implementing strengthening to ensure sustainable and cost-effective solutions are developed and unnecessary strengthening is avoided.

emerald PUBLISHING ice Publishing

Kaloyana Kostova and Craig Giaccio
ISBN 978-1-83549-935-1
https://doi.org/10.1108/978-1-83549-934-420251002

Chapter 2
Materials

General

As stated earlier, the design rules in Annex J of *BS EN 1992-1-1:2023* (BSI, 2023b) were written when harmonised European standards for production and execution were not available, so flexibility was needed in the provisions.

The intent of Annex J is to enable planned development of European Assessment Documents (EADs) to be used (normally cited as *relevant product standards* or *more accurate information available from tests data*) in an effort to ensure that future developments are not prevented by standardisation in *BS EN 1992-1-1:2023*. ISO 10406-1:2015 (ISO, 2015a), ISO 10406-2:2015 (ISO, 2015b) and ISO 10406-3:2019 (ISO, 2019b) provide methods to obtain some relevant design properties, and these have been referenced when available. BS EN 2561:1996 (BSI, 1996) and ISO 527 (ISO, 2019a) also provide methods to determine properties for FRP systems.

The code only caters for strengthening systems and approaches where sufficient databases exist to substantiate code based on the reliability required in the Eurocode standards. This precludes various systems described throughout this book which are in use. These could potentially be covered under future EADs or ETAs.

Partial factors

Provided that all provisions in this standard are complied with, failure mechanisms for the CFRP system arise from either rupture of the CFRP material or bond failure at the interface between the CFRP system and the concrete. Two sets of partial factors for resistance differentiate between the possible mechanisms of failure. These are provided in Table 2.1.

Table 2.1 Partial factors for CFRP systems

Design situation / limit state	Tensile strength		Bond strength
	CFRP strips and bars	In situ lay-up CF sheets	
Designation		γ_f	γ_{BA}
Persistent and transient	1.30	1.40	1.50
Accidental	1.1	1.15	1.15
Serviceability	1.00	1.00	1.00
Fatigue	1.30	1.40	1.50

In situ lay-up CF sheets generally have higher partial factors than CFRP bars and strips. This, in part, reflects the fact that they are impregnated with resin to bond the carbon fibres together on site, hence do not have factory controls on geometry of the complete CFRP product and, therefore, have more variability in resistance.

Material properties

CFRP is a non-ductile linear elastic material which generally has a higher tensile strength compared with ordinary reinforcement steel. Its modulus of elasticity depends on how it is produced and can vary significantly for different systems. The modulus of elasticity of CFRP is also generally low as a ratio to yield strength when compared to ordinary reinforcing steel. A typical stress-strain diagram is shown in Figure 2.1 which illustrates a typical commercially available system alongside ordinary reinforcement (with post yield hardening not considered).

Mechanical properties

The characteristic short-term tensile strength of CFRP, f_{fuk}, will depend on the CFRP strengthening system used and the supplier that produces it. This property should be obtained from a supplier. The material is prone to reductions in strength over time, owing to effects of the environment (such as being in an alkaline environment or exposure to high or low temperatures) and creep under sustained loading. The design value of the tensile strength f_{fud}, therefore, needs to be factored to account for such strength reductions, as is shown in Equation 2.1.

$$f_{fud} = \frac{\eta_f f_{fuk}}{\gamma_f} \tag{2.1}$$

Where

η_f is the reduction applied to the short-term tensile strength to account for environmental conditions and produce the design tensile strength.

γ_f is the partial material factor given in Table 2.1.

Figure 2.1 Typical stress-strain diagram for the CFRP material

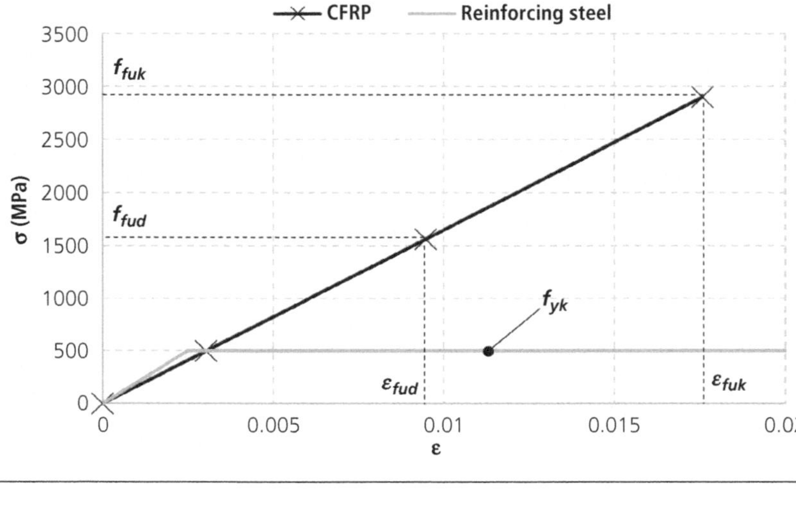

η_f will depend on the production method and it has been shown to vary from 0.65 to 1.0. A value of $\eta_f = 0.7$ is provided as a generally conservative approach. EADs under development will specify more detailed methods for evaluation for producers, which means that the product-specific data will be able to provide recommended values for η_f (refer to *BS EN 1992-1-1:2023*, J.5.3.1(2)). It is recommended that each system is checked once production information is available to ensure that an appropriate value of η_f is used in design. Producers should also be informed of the environment in which the structure is located to ensure that the value of η_f caters for the actual environment of the structure. Durability and environmental considerations are discussed later in this chapter.

The design ultimate strain is determined using the long-term design strength, and the modulus of elasticity of the CFRP in the longitudinal direction of the system, as shown in Equation 2.2.

$$\varepsilon_{fud} = \frac{f_{fud}}{E_f} \tag{2.2}$$

E_f is the mean value of the modulus of elasticity for the CFRP system which should be obtained from supplier's information.

Strength and stiffness

Variability exists in the properties of the fibres resulting from the fibre production process, from the different proportions of fibres and binding resin in a CFRP matrix resulting, and from the different methods available for adding the binding resin to the system. Table 2.2 illustrates the variability in bare fibre properties. ACI 440.2R-17 (ACI, 2017) indicates similar variability properties of laminates that have 40–60% of fibres within the CFRP matrix prevail as those in Table 2.2.

Table 2.2 Ranges of CFRP mechanical properties (fibres only) (Reproduced from fib Bulletin No. 90 (fib, 2019))

Description	Modulus of elasticity [GPa]	Tensile strength [MPa]	Ultimate tensile strain [%]
High strength	215–235	3500–4800	1.4-2.0
Ultra high strength	215–235	4800–6000	2.0-2.3
High modulus	350–500	2500–3100	0.5-0.9
Ultra high modulus	500–700	2100–2400	0.2-0.4

While correlation between strength and stiffness was attempted with the aim of introducing a classification system similar to reinforcing steel into the code during the code preparation process, this was not achieved. The code, therefore, requires input (or specification) of a characteristic short-term tensile strength which will be a specific property.

Limitation on use

BS EN 1992-1-1:2023, J.5.1(3) provides various limits which apply to the design provisions given in the standard. They can be categorised as follows.

- Limits that arise from available testing databases used in the development of the provisions
  - The mean modulus of elasticity of CFRP strips must be in the range of

    $$150000\ \text{MPa} \leq E_f \leq 250000\ \text{MPa}$$

  - The elastic stiffness per unit width of CF sheets must be in the range of

    $$20\ \text{kN/mm} \leq E_f \frac{A_f}{b_f} \leq 400\ \text{kN/mm}$$

  - The total area of all layers of CF sheets in the direction of the applied tension effect must be in the range of

    $$100\ \text{mm}^2 \leq \frac{A_f}{b_f} \leq 1800\ \text{mm}^2$$

- Limits that are included to achieve failure of the concrete material prior to adhesive or interlamellar CFRP material failure
  - Interlaminar shear strength of CFRP strips according to ISO 14130:1997 (ISO, 1997) shall be equal or larger than the adhesive bond strength for any system.
  - Characteristic tensile strength of the adhesive, f_{Atk}, determined in accordance with BS EN 1504-4:2004 (BSI, 2004b) must satisfy $f_{Atk} \geq 14\,\text{MPa}$.

This clause also confirms the requirement that the concrete surface onto which CFRP is bonded is not subject to any existing deterioration in order to ensure that a strengthening system will adhere to the concrete for its intended design life. *BS EN 1992-1-1:2023*, Annex I may be used as a source of finding the types of deterioration that can occur. A site evaluation of the structure being strengthened needs to be undertaken to confirm that the concrete strata onto which the CFRP is adhered to is not deteriorated in any way. This ensures the surface pull-off strength (discussed later) can be generated, and the bond checks are valid.

Cracks that are present where strengthening is applied can result in either surface delamination or crushing of fibres in the FRP matrix. ACI 440.2R (ACI, 2017) recommends that cracks wider than 0.3 mm are pressure-injected with epoxy prior to installation of FRP. ACI 440.2R also recommends that smaller cracks are filled with resin using resin injection in aggressive environments. Further discussion of all matters related to assessment of the suitability of a structure for strengthening and surface preparation are provided in Chapter 10.

Determination of cross-sectional area

The cross-sectional area of the CFRP system is required to determine relevant design parameters and the resistance of members strengthened with CFRP. The cross-sectional area of a CFRP system depends on the type of the CFRP system. Since sheet systems comprise carbon fibres that are woven together, the area of fibres may be different to the area of the system as applied. The code acknowledges this, as follows:

- For sheets, the cross-sectional area of the fibres within a single sheet will be provided by suppliers;
- For strips, the area of each strip may be calculated as the cross-sectional area of the plate or strip.

It may be required that more than one layer of material is used for a strengthening scheme. In this instance, the code requires the following to be considered

- CF sheets.
  - *BS EN 1992-1-1:2023*, J.12.4(1) requires that no more than five layers of CF sheets should be used for flexural or shear strengthening;
  - *BS EN 1992-1-1:2023*, J.12.4(1) requires that no more than ten layers of CF sheets should be used for compression confinement;
  - *BS EN 1992-1-1:2023*, J.5.2(4) requires that where multiple layers of CF sheets are used, the total thickness shall be calculated using Equation 2.4.
- CFRP strips.
  - *BS EN 1992-1-1:2023*, J.12.4(2) requires that no more than two layers of CFRP strips should be used for strengthening;
  - *BS EN 1992-1-1:2023*, J.12.4(2) also requires that the total thickness of CFRP strengthening with CFRP strips $t_f \leq 3\,\text{mm}$.

The cross-sectional area of a CFRP system may be calculated using Equation 2.3.

$$A_f = b_f t_f \tag{2.3}$$

Where

b_f is the sheet or strip width

t_f is the thickness of the CFRP system.

For CFRP strips, t_f is simply the thickness of the strip, or for a multilayered system - the total thickness of all strips.

Where multiple layers of CF sheets are used, the effectiveness of the adhesive to transfer shear between layers reduces. To reflect this, the *BS EN 1992-1-1:2023* defines an effective thickness for CF sheets, which is calculated using Equation 2.4.

$$t_f = n_f^{\,k_f} \frac{A_{f,1}}{b_f} \tag{2.4}$$

Where

n_f is the number of layers of sheets used in strengthening

$$k_f = \begin{cases} 1, & k_f \leq 3 \\ 0.85, & k_f > 3 \end{cases} \tag{2.5}$$

k_f, as shown in Equation 2.5, takes into account the ability of a CFRP strengthening system comprising of multiple layers of CF sheets to transfer shear between each layer in order to function as a single strengthening layer.

$A_{f,1}$ is the cross-sectional area of the fibres of a single layer of a CF sheet, provided by the supplier, for width b_f.

t_f can then be used in Equation 2.3 to determine the total area of the CFRP system comprising multi-layered sheets (A_f)

Surface tensile strength of concrete

The design rules for bond of CFRP and fatigue require the evaluation of a parameter $f_{ctm,surf}$, which can be evaluated from testing of existing structures in accordance with BS EN 1542:1999 (BSI, 1999).

$f_{ctm,surf}$ is the tensile strength of the concrete surface perpendicular to the surface being strengthened. It is this property that is used in *BS EN 1992-1-1:2023* to determine the bond resistance limited by failure of the concrete surface.

As outlined in Chapter 3, *BS EN 1992-1-1:2023*, Formula J.34 provides a method for determining $f_{ctm,surf}$ during the design stage, prior to commencement of work on site, without the need to undertake intrusive testing. *BS EN 1992-1-1:2023*, J.5.3.1(5) includes a requirement to undertake site validation tests of this parameter.'

Durability considerations

It is imperative that the selected CFRP system is appropriate for the environment in which the strengthened structure is located. This is a specialist area, and various references (e.g. fib Bulletin No. 90 (fib, 2019) and TR55 (Concrete Society, 2013) and Al-Mahaidi and Kalfat (2018)) provide good discussions of key points that are to be considered in design. A summary is provided in this section for awareness.

The designer needs to consider all relevant points related to concrete and steel durability, as well as items discussed in this section. These all need to be communicated to the CFRP supplier to ensure that the supplier can provide a system which can resist environmental influences with a view to achieving the required design life. The likelihood of durability issues can, in part, be reduced with adequate surface preparation and protection measures, some of which are outlined in Chapter 10.

In service and application temperature

A notable property of CFRP materials is called the glass transition temperature. It stems from the propensity of the adhesives and resins, used to form the matrix and bond the CFRP strengthening systems to a concrete surface, to soften with elevated temperatures. This is a common phenomenon across all FRP systems. For some systems, temperatures only need to exceed 60°C for softening to occur. The glass transition temperature is the temperature at which a system will begin to display such softening effects. Various methods are available to calculate this temperature, which include BS EN 12614:2004 (BSI, 2004c).

As discussed later in this book, conveying the temperature ranges of the structure being strengthened to a supplier is key to ensuring exceedance of the glass transition temperature does not occur. It is suggested that maximum air and surface temperatures of the structure (in accordance with BS EN 1991-1-5:2003 (BSI, 2003)) are communicated to the supplier, in addition to air temperatures for completeness. Any construction or maintenance operations that will heat the concrete surface also need to be communicated – for example, applying hot rolled asphalt to a bridge deck.

Problems can also arise when applying CFRP in cold temperatures. If not properly addressed, these can result in substandard mixing or application of adhesive which can lead to future durability problems. This is further discussed in Chapter 10 of this book.

Figure 2.2 Illustration of potential sources of moisture ingress (Adapted from fib Bulleting No. 90 (fib, 2019))

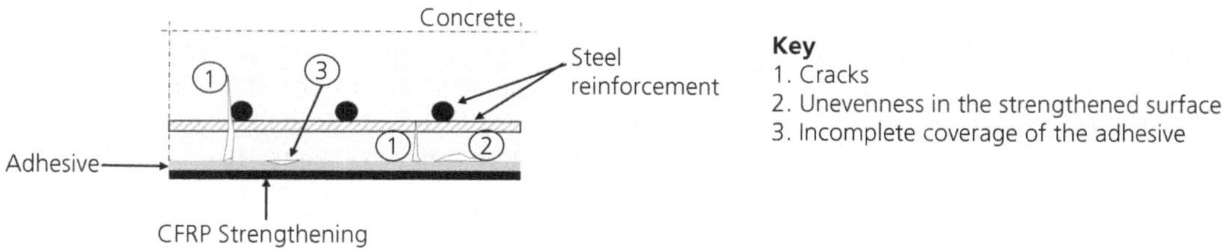

Moisture

FRP products generally have a very good track record of resistance to moisture, which is evidenced by their continued use in boat manufacture. However, there are various places where moisture can be present in a member strengthened with CFRP, examples of which are provided in Figure 2.2.

The resin will absorb trapped water, and this could lead to a reduction in the glass transition temperature, and the resin may stiffen. Trapped water can lead to freeze-thaw effects in areas prone to these. Surface preparation and adhesion of the system can eliminate moisture ingress. This is discussed in more detail in Chapter 10.

Exposure to ultraviolet light (UV)

Exposure to UV can affect the mechanical properties of CFRP. The effect will depend on the manufacture of the system. It is important to specify the external environment and the intended design life to a supplier so they can advise of any issues that might be experienced. UV can also lead to discoloration.

If there is a risk of degradation of mechanical properties, protection systems can be used. Such protection systems include acrylic or polyurethane based paints or coating. Suppliers are best placed to recomment suitable protection products for their systems.

Alkaline and acidic environments

Alkaline and acidic environments can lead to a reduction in properties of a CFRP system, although carbon fibre is noted to perform better than other FRP strengthening systems (refer to ACI440.2R-17 (ACI, 2017) and fib Bulletin No. 90 (fib, 2019)). Selection of the resin will be key for the performance of CFRP systems, and a supplier will need to be informed of such environments.

Saline environments

The use of FRP materials in the boat industry for decades has shown that FRP systems can be very resistant to highly saline environments. However, there is evidence that the resin may experience issues over prolonged periods. Therefore, resins will need to be selected to suit the environment and protection may be required. This will be best specified in conjunction with a supplier.

Freeze-thaw

The topic of impact of freeze-thaw on the performance of CFRP strengthening is researched but limited robust guidance is available for discussion. For reference, fib Bulletin No. 90 (fib, 2019) provides a useful summary of studies that have investigated the matter. One significant consideration for CFRP is the potential for freezing and thawing of trapped water. When water freezes, it expands, which may degrade the CFRP bond over time.

Freeze-thaw can potentially also be more of an issue for an existing structure being strengthened. The structure may not have a concrete specification adequate to resist freeze-thaw effects. It may also have defects which cause water to build up in the concrete. Efforts need to be made to remove these defects when applying the CFRP strengthening system (refer to Chapter 10).

Kaloyana Kostova and Craig Giaccio
ISBN 978-1-83549-935-1
https://doi.org/10.1108/978-1-83549-934-420251003

Chapter 3
Anchorage of adhesively bonded CFRP systems

Introduction

The bond of EBR CFRP strengthening systems frequently governs the resistance of members that are strengthened in flexure or shear. The bond of NSM CFRP strengthening systems sometimes governs the resistance of members that are strengthened in flexure or shear. The bond is also the underlying consideration related to fatigue of EBR CFRP strengthening systems. For this reason, the topic of bond is covered in an early chapter of this book so that examples of flexure and shear strengthening, and fatigue checks can be comprehensively covered in the latter chapters.

Bond resistance of EBR systems

The provisions in *BS EN 1992-1-1:2023* (BSI, 2023b), J.11.1.2 cover the following.

- End anchorage resistance of the CFRP system against an applied tension force in CFRP from adhesion of the CFRP to the concrete.
- Anchorage resistance of the strengthening system between flexural cracks, called intermediate crack debonding resistance.
- End cover separation resistance.
- Shear induced separation resistance.

Bond failure mechanisms of the EBR CFRP system are illustrated in Figure 3.1. Satisfying each condition of the code for each failure mechanism, as well as ensuring CFRP rupture does not occur, results in a compliant EBR CFRP system. A mechanism associated with failure of the adhesive only is not provided in the code as the conditions in *BS EN 1992-1-1:2023*, J.5.1 related to the strength of the adhesive are intended to remove adhesive failure from consideration.

Debonding can also result from poor surface preparation. Issues related to surface preparation are discussed in Chapter 10.

Surface tensile resistance of concrete

Underpinning the design of anchorage of EBR CFRP systems to *BS EN 1992-1-1:2023* is the parameter known as the surface tensile strength of the concrete. It quantifies the resistance of a concrete surface to direct tensile stresses that occur from CFRP strengthening systems, particularly in relation to debonding. An illustration of the test procedure and concrete specimens obtained from the testing of surface tensile strength is shown in Figure 3.2. Prior to *BS EN 1992-1-1:2023* being released, this parameter could only be assessed for the purpose of design using on-site pull-off methods prescribed in BS EN 1542:1999 (BSI, 1999).

Figure 3.1 EBR CFRP failure mechanisms

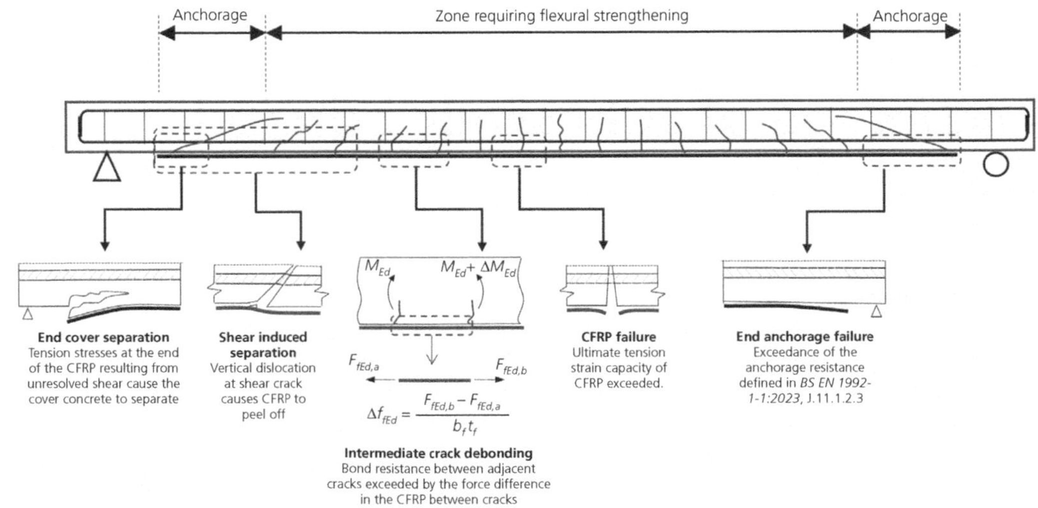

Figure 3.2 Pull-off testing (Photographs courtesy of Concrete Repairs Limited UK)

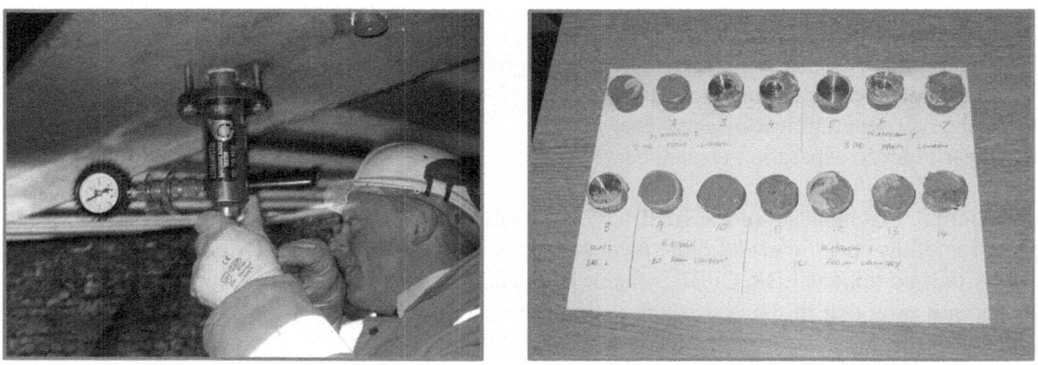

(a) Extraction of pull-off specimen (b) Pull-off specimens

BS EN 1992-1-1:2023, Formula J.34 (Equation 3.1) relates the surface tensile strength for use in design to the mean tensile strength of concrete f_{ctm}. The method sets that the surface tensile strength $f_{ctm,surf}$ is directly proportional to mean tensile strength with a factor $k_{c,surf}$, which depends on the characteristic compressive strength. This factor has been shown in studies to be also influenced by the location of the strengthened surface with respect to the casting method, which is reflected in *BS EN 1992-1-1:2023* (Equation 3.2).

$$f_{ctm,surf} = k_{c,surf} f_{ctm} \tag{3.1}$$

Where

$$k_{c,surf} = \begin{cases} 0.3+0.6\left(\frac{f_{ck}}{60}-0.2\right) & \text{concreting position: top} \\ 0.4+0.5\left(\frac{f_{ck}}{60}-0.2\right) & \text{concreting position: side} \\ 0.6+0.3\left(\frac{f_{ck}}{60}-0.2\right) & \text{concreting position: bottom} \end{cases} \tag{3.2}$$

Where the concrete positions in the case of concrete cast from top are defined in Figure 3.3.

BS EN 1992-1-1:2023, J.5.3(5) provides a requirement to validate the value used in design. This is to ensure that an unconservative value of $f_{ctm,surf}$ is not used in design.

Anchorage resistance and length (refined method)

As described in fib Bulletin No. 90 (fib, 2019), if concrete deformation is ignored at the EBR CFRP interface, the basic anchorage length of CFRP can be determined according to the differential equation for bond shown in Equation 3.3.

$$\frac{d^2s}{dx^2} - \frac{p_f}{E_f\,A_f}\tau_b(s) = 0 \tag{3.3}$$

Where

s is the slip of the EBR CFRP system

p_f is the bonded perimeter of the CFRP

E_f is the mean modulus of elasticity of the CFRP system

A_f is the area of CFRP.

Figure 3.3 Definition of concrete casting positions for determining $k_{c,surf}$

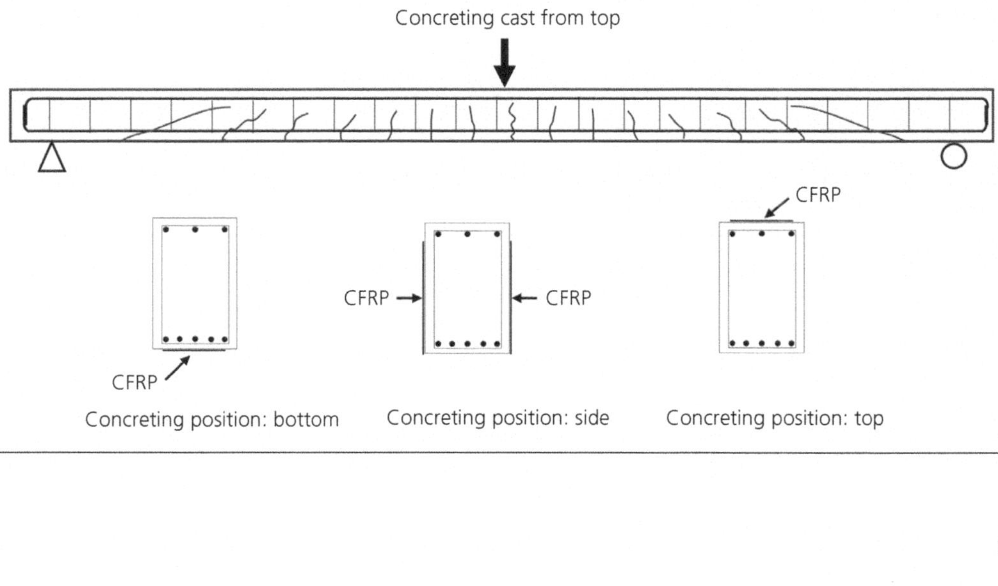

$\tau_b(s)$ is the bond-slip law describing the relationship between the shear stress between the CFRP and concrete substrate, τ_b, and the slip required to generate this shear stress, s. A bilinear shear slip relationship is commonly used, as illustrated in Figure 3.4.

As outlined in the Background Document for FprEN 1992-1-1 (European Committee for Standardisation, 2023), a closed-form solution to the differential equation defining bond (Equation 3.3) is not possible with a bilinear relationship between the shear stress and slip. A solution is, though, possible with a linear softening relationship, where s_{f1k} in Figure 3.4 is zero.

The characteristic value of the anchorage length, $l_{bf,max,k}$, required to generate the maximum bond stress, $f_{bfk,max}$, is defined in Equation 3.4. This expression is obtained by solving the bond-slip equation (Equation 3.3). A modification has been introduced to cater for the fact that implementing the simplification for the bond-slip law (i.e. taking $s_{f1k} = 0$) underestimates the anchorage length needed to generate $f_{bfk,max}$ (refer to Background Document to FprEN 1992-1-1 (European Committee for Standardisation, 2023)).

$$l_{bf,max,k} = \frac{2}{k_{sys,b3}} \sqrt{\frac{E_f \, t_f \, s_{f0k}}{\tau_{f1k}}} \tag{3.4}$$

The characteristic maximum value of shear stress at the interface between EBR CFRP is shown in Equation 3.5 (*BS EN 1992-1-1:2023*, Formula J.42).

$$\tau_{f1k} = 0.37 k_{sys,b1} \sqrt{f_{cm} \, f_{ctm,surf}} \tag{3.5}$$

Equation 3.6 (*BS EN 1992-1-1:2023*, Formula J.41) defines the characteristic value of slip at τ_{f1k}. This equation effectively sets the characteristic slip to $s_{f0k} = 0.2\,\text{mm}$, but also includes the term $k_{sys,b2}$, which may be used to ensure that characteristic values used in design align with production data provided by suppliers.

$$s_{f0k} = 0.2 k_{sys,b2} \tag{3.6}$$

The characteristic maximum resistance of an end anchorage, $f_{bfk,max}$, is defined as shown in Equation 3.7 (*BS EN 1992-1-1:2023*, Formula J.40).

$$f_{bfk,max} = \sqrt{\frac{E_f \, \tau_{f1k} \, s_{f0k}}{t_f}} \tag{3.7}$$

Figure 3.4 General form of bond-slip relationship for EBR CFRP systems (Reproduced from fib Bulletin No. 90 (fib, 2019) using characteristic values)

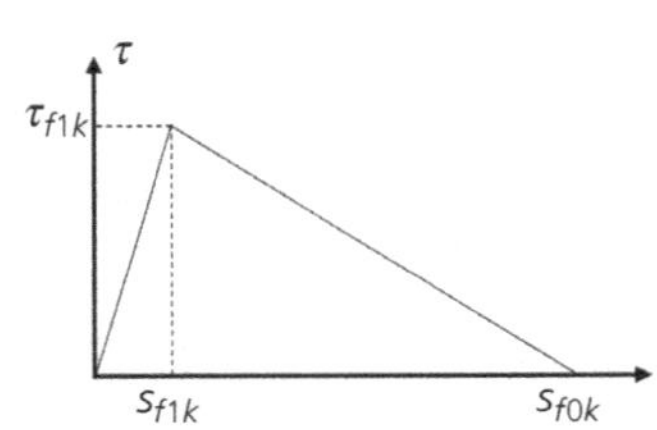

Equation 3.5, Equation 3.6 and Equation 3.4 include terms of the form $k_{sys,bx}$. These terms allow for suppliers to modify anchorage parameters based on their systems. The intention of the code, when being drafted, was to ensure that determination of these values is undertaken in accordance with European Assessment Documents under preparation at the time the code was being written. These European Assessment Documents were not available at the time this book was finalised.

The design CFRP anchorage resistance f_{bfRd} for a given anchorage of length l_{bf} is defined in Equation 3.8.

$$f_{bfRd} = \frac{\sqrt{\eta_{cc} k_{tc} k_{tt}}}{\gamma_{BA}} \begin{cases} f_{bfk,max} \dfrac{l_{bf}}{l_{bf,max,k}} \left(2 - \dfrac{l_{bf}}{l_{bf,max,k}} \right) \text{where } l_{bf} < l_{bf,max,k} \\ f_{bfk,max} \text{ where } l_{bf} \geq l_{bf,max,k} \end{cases} \quad (3.8)$$

Simplified method for determining anchorage resistance and length

In drafting the code provisions, the simplified method provided in *BS EN 1992-1-1:2023*, J.11.2.2 aimed to remove any reliance on production data from suppliers during design – that is, setting $k_{sys,bx} = 1.0$. If one assumes recommended values of k_{tc} and k_{tt} (i.e. $k_{tc} = 0.85$ and $k_{tt} = 0.7$), the method will generally produce slightly conservative results (i.e. underestimating f_{bfRd} and overestimating $l_{bfk,max}$) up to a characteristic concrete strength of approximately 50 MPa. Beyond this, the method will overestimate f_{bfRd} when compared to the refined method. Care should be taken when applying this method to higher-strength concrete.

Intermediate crack debonding

Anchorage resistance between cracks is limited by the discontinuity in the concrete substrate onto which the CFRP bonds resulting from crack formation. Debonding between cracks is known as intermediate crack debonding. The EBR system design needs to ensure sufficient bond of the system between cracks, as illustrated in Figure 3.5. *BS EN 1992-1-1:2023* requires that these checks are undertaken at a minimum crack spacing, which is calculated as per Equation 3.9 (*BS EN 1992-1-1:2023*, Formula J.44).

The formula for minimum crack spacing is of a different format to the code formulae for crack spacing. This was simply a result of the fact that calibration of the models had been undertaken with the code method.

$$s_{cr,min} = 1.5 \frac{M_{cr}}{0.85 h F_{bsm}} \quad (3.9)$$

Figure 3.5 Illustration of an element of concrete strengthened with EBR CFRP between cracks

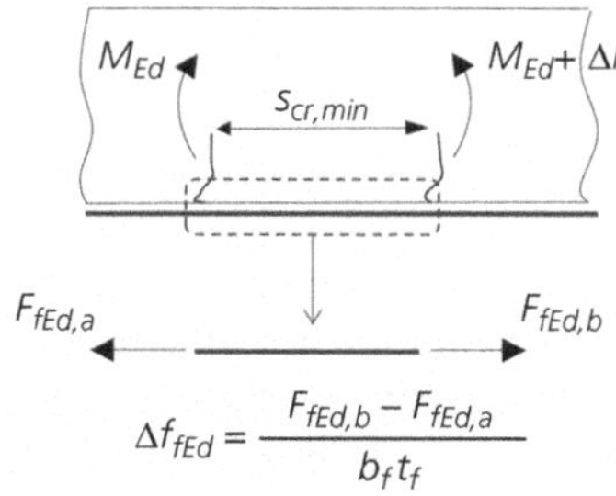

Where F_{bsm} is the bond strength per unit length of reinforcement, calculated in accordance with *BS EN 1992-1-1:2023*, Formulae J.45 and J.46 in the code.

For compliance with the code requirements, Equation 3.10 must hold.

$$\Delta f_{fRd} \geq \Delta f_{fEd} \tag{3.10}$$

Where

Δf_{fEd} is defined in Figure 3.5.

The intermediate crack debonding resistance, Δf_{fRd}, is defined in Equation 3.11 (*BS EN 1992-1-1:2023*, Formula J.47).

$$\Delta f_{fRd} = \frac{\sqrt{\eta_{cc} k_{tc} k_{tt}}\, \Delta f_{fk,B} + \Delta f_{fk,F} + \Delta f_{fk,C}}{\gamma_{BA}} \tag{3.11}$$

Δf_{fRd} comprises three separate components, as described below. As explained in Background Document to FprEN 1992-1-1 (European Committee for Standardisation, 2023), the bond component is determined using the bond-slip law with a bilinear approach, and additional contributions are included from bond friction and a clamping force from the curvature of a member.

The three components of intermediate crack debonding resistance comprise the following.

- A component of resistance from the strength of the adhesive bond between the CFRP and concrete substrate, $\Delta f_{fk,B}$, defined in Equation 3.12 (*BS EN 1992-1-1:2023*, Formula J.48).
- A component of resistance from friction between the CFRP and concrete substrate, $\Delta f_{fk,F}$, defined in Equation 3.13 (*BS EN 1992-1-1:2023*, Formula J.49).
- A component of resistance from clamping owing to curvature of the beam, $\Delta f_{fk,C}$, defined in Equation 3.14 (*BS EN 1992-1-1:2023*, Formula J.50).

$$\Delta f_{fk,B} = 0.84 k_{sys,b1} \sqrt{f_{cm} f_{ctm,surf}}\, \frac{\sqrt{s_{cr,min}}}{t_f} \tag{3.12}$$

$$\Delta f_{fk,F} = f_{cm}^{-0.9} \frac{s_{cr,min}^{\frac{4}{3}}}{t_f} \tag{3.13}$$

$$\Delta f_{fk,C} = \frac{\kappa_h}{\max(100\,\text{mm}, h)} \frac{s_{cr,min}^{\frac{1}{3}}}{t_f} \tag{3.14}$$

The values of κ_h recommended in *BS EN 1992-1-1:2023* suggest that the clamping force only applies to reinforced concrete beams. TR55 (Concrete Society, 2013) indicates that a hogging curvature (which can result from deformation or from the cast concrete profile) can be counterproductive and reduce the bond resistance by generating local transverse pull-off forces that result from the curvature. While the beneficial effect from clamping has been demonstrated in literature (e.g. fib Bulletin No. 90 (fib, 2019)), it is recommended that caution is taken when considering this effect as it should only be relied on where a sagging curvature can be demonstrated to generate this additional clamping force. As stated in TR55 (Concrete Society, 2013), it is recommended that, where a hogging curvature is evident, specialist advice should be sought. For reference, ACI 440.2R-17 (ACI, 2017) recommends that the surface curvature of concave surfaces should not exceed 5 mm per metre, and TR55 (Concrete Society, 2013) recommends the same for plate-based systems and 3 mm per metre for CF sheets. More details on this are provided in Chapter 10.

The opening statement of *BS EN 1992-1-1:2023*, J.11.1.2.3(1) suggests that intermediate crack debonding checks only need to be undertaken when the refined method of determining the anchorage length is used – that is, not when the simplified method is used. Given the illustration provided in this book of how close the two methods are, it is considered appropriate to undertake this check to ensure that the CFRP strengthening system is anchored throughout its full length.

The theoretical approaches described above all assume a well-prepared even surface onto which the CFRP is bonded. If this is not the case, there may be additional local forces generated which can increase the applied debonding forces. The preparation of the surface is explained in more detail in Chapter 10. It is imperative that this preparation is undertaken to ensure that the theoretical approaches described here are validated by site works.

End cover separation

In certain circumstances, it is possible for the cover concrete to separate at the end of an EBR CFRP strengthening system. This phenomenon is illustrated in Figure 3.6. Cover separation at the end of the anchorage zone occurs due to unresolved tension forces that arise in the shear mechanism illustrated in Figure 3.7.

Figure 3.7 illustrates that the shear reinforcement forces can resolve themselves through the longitudinal reinforcement and concrete struts. CFRP, similarly, utilises the shear reinforcement to resolve internal forces. However, at the end of the anchorage area, a potential unresolved tension force can arise which causes end cover separation.

Figure 3.6 End cover separation failure mode

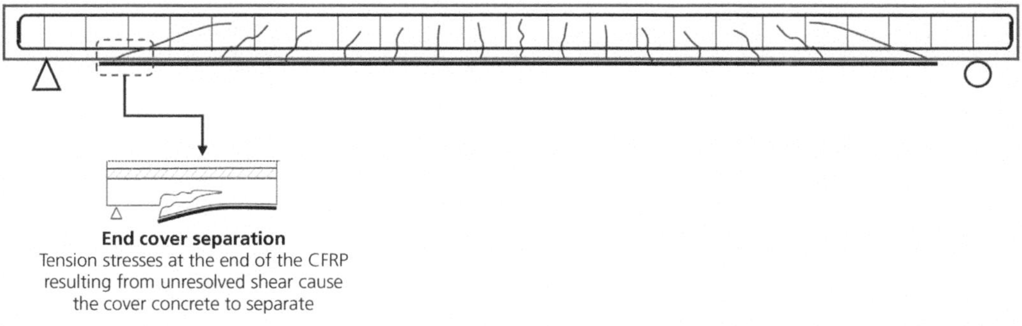

Figure 3.7 Illustration of force system causing end cover separation (Copied from Background Document to FprEN 1992-1-1 (European Committee for Standardisation, 2023), page 685)

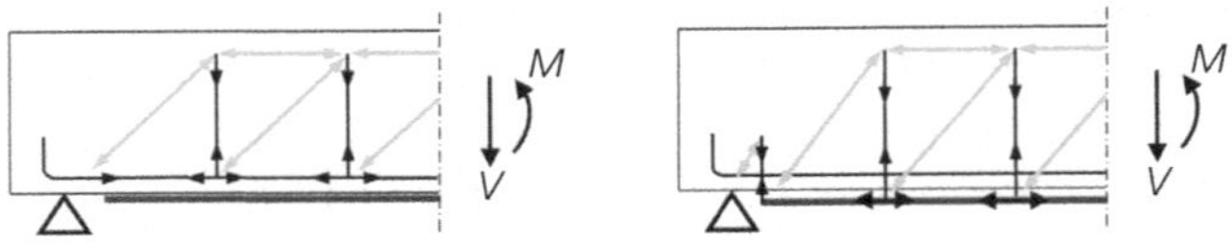

(a) Internal force system

(b) Internal force system including EBR CFRP

To ascertain if shear strengthening is needed to prevent end cover separation, a shear check is undertaken. The design shear force at the end of the strengthening system, V_{Ed}, is checked against the condition in Equation 3.15 (*BS EN 1992-1-1:2023*, Formula J.51).

$$V_{Ed} \leq V_{Rd,cfE} = \left(0.11 + 2.2\frac{(100\rho_l)^{0.15}}{(a_{fE})^{0.36}}\right)(100\rho_l f_{ck})^{\frac{1}{3}} b_w d_{st} \tag{3.15}$$

Where

ρ_l is the ratio of longitudinal tension reinforcement to the gross concrete area

d_{st} is the effective depth of the section

b_w is the minimum width of the cross section between tension and compression chords and the neutral axis.

The term a_{fE} is the distance between the end of the CFRP strengthening and the nearest point of zero moment (including a point of contraflexure in a hogging zone), including allowance for the shift term. It is defined in Figure J.6 in *BS EN 1992-1-1:2023*.

Where the condition in Equation 3.15 is not satisfied, an additional CFRP strip or sheet applied as a stirrup is required to resist the vertical component of a force that would be generated at the nodal connection between the CFRP strengthening and the shear stirrup, when the force in the longitudinal CFRP is equal to the bond resistance F_{bfRd}, as shown in Figure 3.7. This CFRP design stirrup force is designated F_{st} in Figure 3.8.

Mathematically, F_{st} can be determined as shown in Equation 3.16.

$$F_{st} = F_{bfRd} tan\theta \tag{3.16}$$

Where

θ is the strut angle for determining shear resistance, which is illustrated in Figure 3.8.

BS EN 1992-1-1:2023, Formula J.52 expresses this in terms of stress. The resistance requirement can be rewritten as in Equation 3.17.

$$\frac{F_{st}}{A_f} = \frac{F_{bfRd} tan\theta}{A_f} \leq f_{fwd} \tag{3.17}$$

Figure 3.8 Illustration of vertical component of force to be resisted by additional stirrup for end cover separation

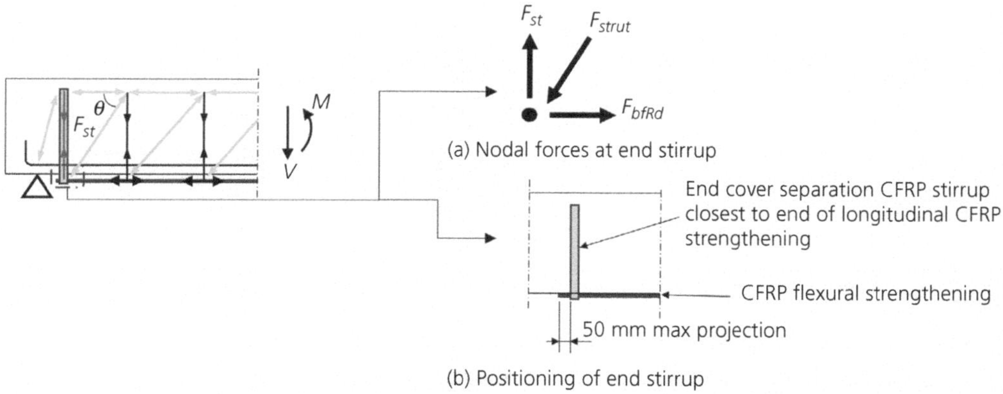

Where

A_f in Equation 3.17 is the area of the CFRP shear strengthening (account for two legs where provided)

f_{fwd} is the design strength of the strengthening system, which can be calculated using *BS EN 1992-1-1:2023*, Formulae J.18, J.19, J.20 or J.23 from the shear section of the code. These terms are described later in this book.

Also shown in Figure 3.8 is a maximum distance of 50 mm between the end of the flexural strengthening and the end cover separation stirrup, which needs to be provided in detailing. This distance is important to ensure that the end cover separation stirrup is able to arrest the failure mechanism effectively.

Where it is not possible to provide a shear link (e.g. when strengthening the soffit of a slab) and an alternative anchorage point cannot be found, such that the requirements of Equation 3.15 are met, then a mechanical anchorage system may be required. This is not covered in *BS EN 1992-1-1:2023*, and specialist and/or supplier advice may be needed.

Shear induced separation

As illustrated in Figure 3.9, shear cracking can result in a vertical separation of the tension surface of a member subject to flexure. This vertical separation can cause debonding of the CFRP strengthening system.

To cater for the issue, Equation 3.18 (*BS EN 1992-1-1:2023*, Formula J.53) imposes a limit on stress in the steel stirrups as a means of limiting vertical displacement at a shear crack. As the code applies to existing structures, a limit is set for plain round bars as well as deformed bars. Once this limit is exceeded, additional shear strengthening is needed to limit the vertical displacement at a shear crack (refer to Equation 3.23).

$$\frac{\tau_{Ed}\sigma_{swd}}{\tau_{Rd}} \leq \begin{cases} 75 \text{ MPa for ribbed bars} \\ 25 \text{ MPa for plain steel bars} \end{cases} \qquad (3.18)$$

Figure 3.9 Illustration of vertical dislocation at shear cracks (Adapted from Background Document to FprEN 1992-1-1 (European Committee for Standardisation, 2023), page 686)

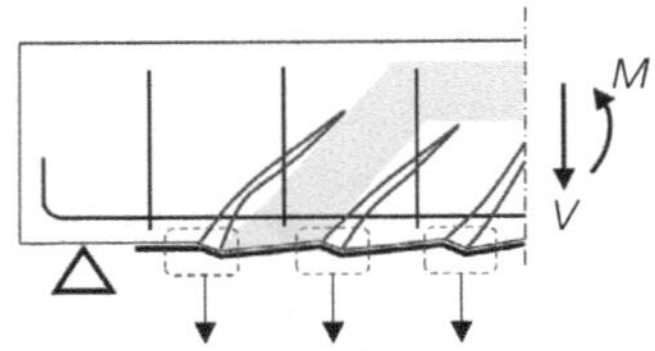

Where

τ_{Ed} is the maximum design shear stress within the strengthened section

τ_{Rd} is the corresponding shear resistance within the strengthened section.

The code refers to *BS EN 1992-1-1:2023*, Formula 8.62 for determining the value of τ_{Rd}. However, this value is only for short shear spans, and should only be used in such applications. Comparison with fib Bulletin No. 90 (fib, 2019) indicates that the intended value here is the upper limit in Formula 8.62, which is the crushing strength of the compression strut, akin to $V_{Rd,max}$ in BS EN 1992-1-1:2004 (BSI, 2004a).

σ_{swd} is the tension stress in the shear links. Where the checks are to be undertaken outside of the region close to the support, this stress can simply be determined by equally distributing the applied shear force V_{Ed} into each leg of each stirrup that crosses the shear crack at the strut angle θ, as shown in Equation 3.19.

$$\sigma_{swd} = \frac{V_{Ed}}{\frac{A_{sw}}{s} z(\cot\theta + \cot\alpha)\sin\alpha} \tag{3.19}$$

The code indicates that this can also be calculated from formulae in the shear section using Equation 3.20 (*BS EN 1992-1-1:2023*, Formula 8.63). Formula 8.63 applies only for members with short shear spans and should only be applied in such applications.

$$\sigma_{swd} = E_s \left[(\varepsilon_x + 0.001) \frac{(cot\theta + cot\alpha_w)^2}{1 + cot^2\alpha_w} - 0.001 \right] \leq f_{ywd} \tag{3.20}$$

Where

E_s is the modulus of elasticity of the shear reinforcement

ε_x is the strain in the longitudinal reinforcement

θ is the strut angle used for assessment of shear resistance τ_{Rd}

α_w is the angle of the shear links measured from the longitudinal axis of the member.

Shear links will commonly be perpendicular to the member longitudinal axis, giving $\alpha_w = 90^o$. This means that Equation 3.20 will become Equation 3.21.

$$\sigma_{swd} = E_s \left[(\varepsilon_x + 0.001) cot^2\theta - 0.001 \right] \leq f_{ywd} \tag{3.21}$$

ε_x is the average of the strains in the top and bottom chord of the member in shear at the section being assessed for shear induced crack separation. Combining *BS EN 1992-1-1:2023*, Formulae 8.56 to 8.62 produces Equation 3.22 for determining this parameter.

$$\varepsilon_x = \frac{\dfrac{\left[\dfrac{M_{Ed}}{z} + \dfrac{\left(|V_{Ed}|cot\theta + N_{Ed}\right)}{2}\right]}{A_{st}E_s} + \dfrac{-\left[\dfrac{M_{Ed}}{z} - \dfrac{\left(|V_{Ed}|cot\theta + N_{Ed}\right)}{2}\right]}{A_{cc}E_c}}{2} \qquad (3.22)$$

The sign conventions in Equation 3.22 (taken from *BS EN 1992-1-1:2023*, Formula 8.47 and 8.48) are such that tension is taken as a positive force and compression is negative. This needs to be input correctly into the equation to ensure the correct strain is obtained.

The UK National Annex or Published Document will likely clarify that the depth of concrete to be used in calculating the area in compression, A_{cc} in Equation 3.22, should be taken as depth of concrete in compression above the neutral axis determined using a rectangular stress block (x_{sb} as defined in Section 3 of *BS EN 1992-1-1:2023*).

Where provisions of Equation 3.18 are not satisfied, additional shear strength is required. Equation 3.23 provides the requirement for determining action effects for shear in this case. Design resistance of a CFRP system to resist this shear is determined using shear provisions stated in *BS EN 1992-1-1:2023*, J.8, and described later in this book.

$$\tau_{Ed,f} = max\begin{cases} \dfrac{E_f A_f}{E_f A_f + E_s A_s}\tau_{Ed} \\ \tau_{Ed} - \tau_{Rd,sy} \end{cases} \qquad (3.23)$$

The reliance of the shear induced separation check on σ_{swd} implies that a member that does not require steel shear reinforcement is not prone to shear induced separation. The code does not though, cover the case of a member without steel shear reinforcement requiring strengthening for shear and flexure. It is suggested that in this case, σ_{swd} in Equation 3.18 may be replaced with Equation 3.24.

$$\sigma_{swd} = \sigma_f \frac{E_s}{E_f} \qquad (3.24)$$

If shear strengthening cannot be provided to resolve a shear induced separation, there is no solution available in the code that can be implemented. Specialist and/or supplier advice should be sought in this instance.

End anchorage (curtailment requirements) for a member strengthened in flexure

BS EN 1992-1-1:2023, J.11.1.2.2 outlines provisions for curtailing a CFRP EBR system. It provides two cases, one which is generally relevant to strengthening members in flexure (e.g. beams) and another which is more applicable to local strengthening (e.g. strut and tie). Curtailment for shear is addressed within the shear provisions described in Chapter 4.

For strengthening members in flexure, curtailment can be undertaken using a similar approach to reinforcement. Curtailment can occur in the cracked part of a member, but an additional longitudinal force in the CFRP in beams with shear reinforcing, owing to the longitudinal component of the diagonal shear force, is taken into account by using the shift rule described in *BS EN 1992-1-1:2023*, 12.3.2. This is shown illustratively in Figure 3.10.

Determining the moment shift relies on finding the inner lever arm of internal forces for shear design, z, defined in Figure 3.11. In this figure, the lever arm is shown to be the distance between the lines of action of the net compression and tension forces. The presence of CFRP means that the CFRP material will also attract tension force and, therefore, should be considered in determining the lever arm z. Literature is inconclusive about how this should

Figure 3.10 Curtailment of EBR CFRP systems

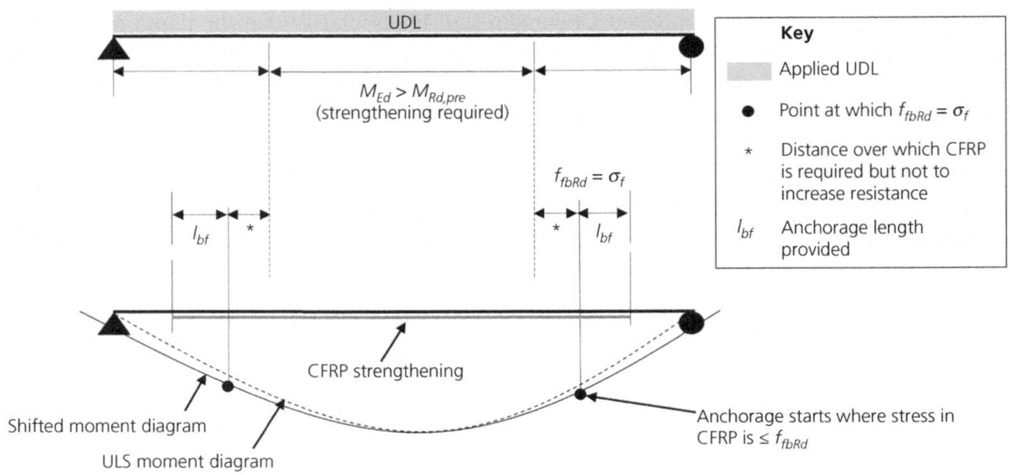

Figure 3.11 Definition of inner shear lever arm (Taken from *BS EN 1992-1-1:2023*, Figure 8.9)

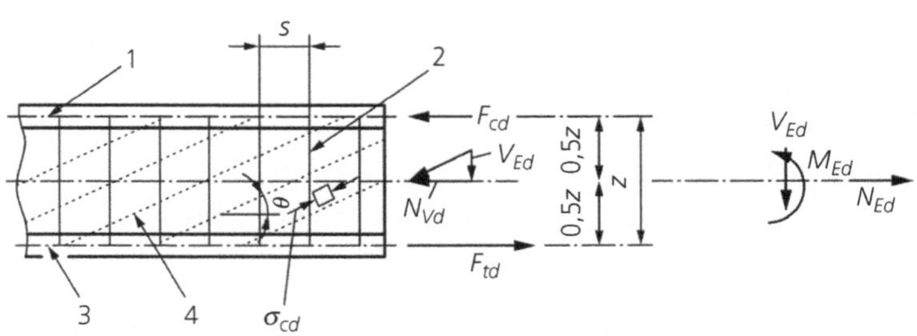

Key
1 Axis of compression chord
2 Shear reinforcement
3 Axis of tension chord
4 Struts (compression field)

be adapted to the case where CFRP is present. The examples in this book assess two methods which are listed below.

(a) Include the CFRP reinforcement in the calculation of the net value and position of the tension force using the modular ratio as appropriate to position the force.
(b) Use a detailed iterative method to assess the full cross-sectional stress state and enable z to be determined.

If the weighted ratio of CFRP stiffness to steel stiffness is high, the inner lever arm can increase. Using an unconservative approach to determine z and, therefore, a_l could mean that the point at which the strengthening system is curtailed is such that the shear at the point of curtailment could be underestimated, leading to an unconservative assessment of end cover separation. The end cover separation failure mechanism is discussed in this chapter.

As shown in Figure 3.10, the curtailment cannot commence until the force in the CFRP is less than the design anchorage resistance calculated using *BS EN 1992-1-1:2023*, J.11.1.1.2 or J.11.1.1.3. Anchorage of CFRP in cracked zones of members strengthened in flexure needs to be such that the bond resistance can be generated between cracks spaced at $s_{cr,min}$. To ensure that this is achieved, it is recommended that where curtailment of flexural strengthening occurs in the cracked region of a member, a minimum anchorage length of $2s_{cr,min}$ is provided. A designer may also wish to provide the full anchorage length of $l_{bf,max,k}$ in a cracked region if this distance is greater than $s_{cr,min}$. As an alternative, as documented in fib90 (2019), there is the possibility to provide CFRP strengthening for the entire length of the cracked member, and provide the anchorage length of l_{bf} outside of the cracked region. In some cases, the latter can achieve the lowest overall CFRP length.

For local strengthening applications, curtailment of an EBR CFRP system should extend a distance h (the member depth) beyond the length $l_{bf,max,k}$, calculated in *BS EN 1992-1-1:2023*, Formula J.3.

fib Bulletin No. 90 (fib, 2019) recommends that where strengthening is undertaken to provide compliance with serviceability limit state (SLS) requirements, the strengthening should be extended throughout the full extent of the cracked section of a member in flexure.

Bond resistance of NSM strips

The bond of EBR NSM systems sometimes limits capacity of flexural and shear strengthening systems. However, the bond of NSM systems is far more effective than EBR systems, and much more resistance is generated from the use of the system per unit length of strengthening application.

BS EN 1992-1-1:2023 deals with the bond of NSM in a fairly straightforward manner. The provisions practically cover only NSM strips, with a note indicating that CFRP rods and bars cannot be designed as NSM strengthening with the provisions in the standard. This limit applies due to a preferential perimeter to cross section ratio that allows better use of the strength of the FRP material, which is described in Kalupahana *et al.* (2013).

The method in the standard distinguishes between a bond failure in the concrete and failure of the adhesive, using single provisions for each condition for strips through the use of the parameter

τ_{bAd}. The bond resistance of NSM strips for a given anchorage length l_{bf} is then determined using Equation 3.25.

$$F_{bfRd} = \begin{cases} 0.95\, b_f \tau_{bAd}\, a_r^{\frac{1}{4}} l_{bf} \left(0.4 - 0.0015 l_{bf}\right) \text{for } l_{bf} \leq 115\,\text{mm} \\ 0.95\, b_f \tau_{bAd}\, a_r^{\frac{1}{4}} \left(26.2 + 0.065\, tanh\left(\frac{a_r}{70}\right)\left(l_{bf} - 115\right)\right) \text{for } l_{bf} > 115\,\text{mm} \end{cases} \quad (3.25)$$

Where, as shown in Equation 3.26

$$\tau_{bAd} = \frac{1}{\gamma_{BA}} \begin{cases} \tau_{bAk}\alpha_{bA} \\ \tau_{bck}\alpha_{bc} \end{cases} \quad (3.26)$$

The adhesive failure is checked using the term $\tau_{bAk}\alpha_{bA}$, where τ_{bAk} is determined in accordance with Equation 3.27.

$$\tau_{bAk} = 0.6\sqrt{\left(2 f_{Atk} - 2\sqrt{f_{Atk}^2 + f_{Atk} f_{Ack}} + f_{Ack}\right) f_{Atk}} \quad (3.27)$$

Where f_{Atk} and f_{Ack} are characteristic strengths of the adhesive in tension and compression, respectively, and are defined in *BS EN 1992-1-1:2023*, J.5.

While the recommended value of α_{bA} is $\alpha_{bA} = 0.5$, the code permits a more accurate assessment of this value to be undertaken. This will require a future European Technical Assessment (ETA) to be available for this assessment based on European Assessment Documents (EADs) currently under preparation.

The concrete failure is checked using the term $\tau_{bck}\alpha_{bc}$ as is shown in Equations 3.28 and 3.29.

$$\tau_{bck} = 4.5\sqrt{f_{cm}} \quad (3.28)$$

$$\alpha_{bc} = \left(\eta_{cc} k_{tc} k_{tt}\right)^{0.5} \quad (3.29)$$

The code also permits a more accurate assessment of α_{bc} to be undertaken. This will again require a future ETA to be available for this assessment based on EADs currently under preparation.

a_r is the distance of the longitudinal axis of the strip to the nearest free edge. The value for a_r for use in Equation 3.25 is limited by $a_r \leq 150$ mm.

The code provisions apply only for NSM CFRP strips with a width in the range of 10 mm $\leq b_f \leq$ 30 mm and thickness in the range of 1 mm $\leq t_f \leq$ 3 mm. Where CFRP rods or bars are used, the bond behaviour is different and additional tests are required to validate the performance. fib Bulletin No. 90 (fib, 2019) provides additional provisions for bars and rods as does TR55 (2013). In the interim, supplier's advice should also be sought if CFRP bars or rods are to be used.

End cover separation in accordance with *BS EN 1992-1-1:2023*, J.11.1.2.4 and shear induced separation in accordance with *BS EN 1992-1-1:2023*, J.11.1.2.5 should also be checked for NSM

CFRP strengthening systems. Despite the code suggesting that only one of the shear induced separation checks applies, the future British National Annex to *BS EN 1992-1-1:2023*, or the accompanying Published Document, is likely to suggest that the full shear induced separation check is undertaken.

In all cases, the shift of the tension envelope in accordance with *BS EN 1992-1-1:2023*, J.12 needs to be taken into account when determining tension forces in the NSM CFRP system.

Kaloyana Kostova and Craig Giaccio
ISBN 978-1-83549-935-1
https://doi.org/10.1108/978-1-83549-934-420251004

Chapter 4
Shear

Introduction

BS EN 1992-1-1:2023 (BSI, 2023b), J.8 covers the majority of ultimate limit state (ULS) items given in the standard. Some specific exclusions are made for use of ABR CFRP strengthening for ULS – for example, strengthening for punching shear. These were excluded in preparation of the code as the database of experimental work was not sufficient to ensure robust code provisions with the required level of reliability could be given in this standard.

The principles for shear strengthening beams are required to undertake full flexural strengthening checks, including all bond analysis. This book therefore presents the shear strengthening chapter first, ahead of the chapters covering flexural, axial and other ULS considerations.

Figure 4.1 (*BS EN 1992-1-1:2023*, Figure J.3) shows that the shear strengthening systems covered by the standard are limited to EBR strips or sheets. This is effectively web strengthening with three- or four-sided wrapped EBR systems, primarily for rectangular sections. NSM strengthening can be used for shear strengthening, but no such provisions are given in the standard. Publications such as TR55 (Concrete Society, 2013) would need to be consulted for such an application. Other shear strengthening systems such as web strengthening of beams only (without wrapping the soffit), or deep embedded rod/bars, are also not covered in the standard.

Shear strengthening of members not requiring shear reinforcement

Shear strengthening of members that do not require any shear reinforcement is covered by *BS EN 1992-1-1:2023*, J.8.2.2. As stated in Clause 8.2.1, shear reinforcement is not needed when the shear resistance of concrete, $\tau_{Rd,c}$, alone is sufficient to resist acting shear forces. $\tau_{Rd,c}$ is calculated according to *BS EN 1992-1-1:2023*, 8.2.2, and provisions of *BS EN 1992-1-1:2023*, Annex I may be used to supplement this section. When applying these provisions, longitudinal CFRP that can be applied for flexural strengthening is not to be used to determine the area of longitudinal reinforcement.

Application of the shift rule in *BS EN 1992-1-1:2023*, 12.3.2 negates the need to consider the additional force in the CFRP in the anchorage zone for flexural strengthening.

Shear strengthening of members with shear reinforcement

As mentioned above, shear strengthening systems covered by provisions stated in the standard are shown in Figure 4.1. These are all externally bonded systems. Provisions for strengthening systems that utilise mechanical anchorages have not been included in this standard as the database of experimental results was not sufficient at the time of preparing the standard to ensure the required reliability of strengthening systems. Future European Assessment Documents (EADs) may cover other shear strengthening systems.

Other limits in the use of Annex J are also included in *BS EN 1992-1-1:2023*, J.8.2.3(2) – (3). These provisions are an acknowledgement of limits in the databases available for developing the design equations.

Figure 4.1 Types of EBR CFRP shear strengthening covered by *BS EN 1992-1-1:2023* (Reproduced from *BS EN 1992-1-1:2023*)

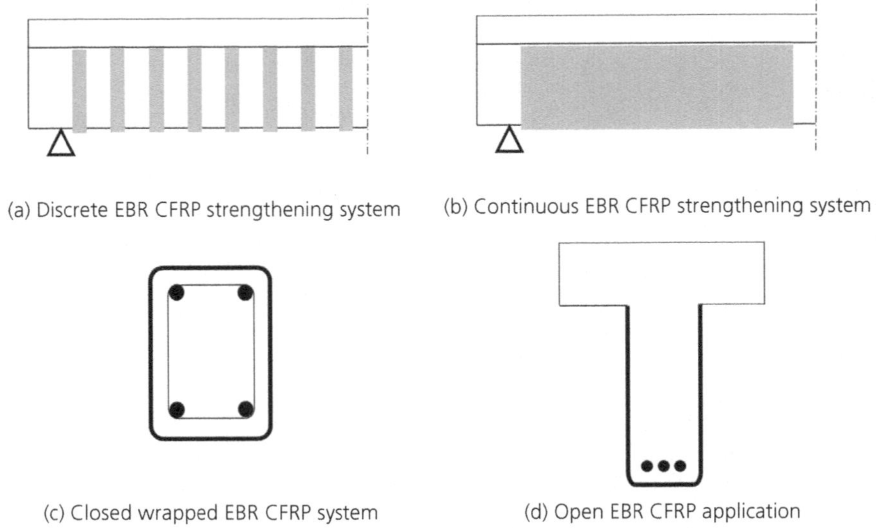

Shear resistance of a member

The shear resistance of a member strengthened with a CFRP system in *BS EN 1992-1-1:2023* is determined using the superposition of the resistance of the concrete member and that of the CFRP system, as is commonly applied in members with steel shear reinforcement. This is shown in Equation 4.1 (*BS EN 1992-1-1:2023*, Formula J.15).

$$\tau_{Rd,CFRP} = \tau_{Rd} + \tau_{Rd,f} \leq 0.5\nu f_{cd} \quad (4.1)$$

Where

τ_{Rd} is the shear resistance calculated in accordance with *BS EN 1992-1-1:2023*, 8.2.2 or 8.2.3, supplemented by *BS EN 1992-1-1:2023*, Annex I where appropriate, as is shown in Equation 4.2.

$$\tau_{Rd,f} = \frac{A_f}{s_f}\frac{f_{fwd}}{b_w}\left(cot\theta + cot\alpha_f\right) \quad (4.2)$$

Where

f_{fwd} is the design shear strength of the strengthening system defined in Table 4.1

α_f is the angle between the fibres in the strengthening system and the longitudinal axis of the beam as shown in Figure 4.2

A_f is the total area of CFRP defined in Equation 2.3 or provided by the supplier.

θ is the strut angle for determining shear resistance.

$0.5\nu f_{cd}$ is the upper limit of shear strength corresponding to strut compression failure. The shear resistance of the member with CFRP strengthening cannot exceed this upper limit.

Table 4.1 Definition of f_{fwd} (continued on next page)

Type of System	Design Shear Resistance of Strengthening System (f_{fwd})	Comments
Fully wrapped CFRP system 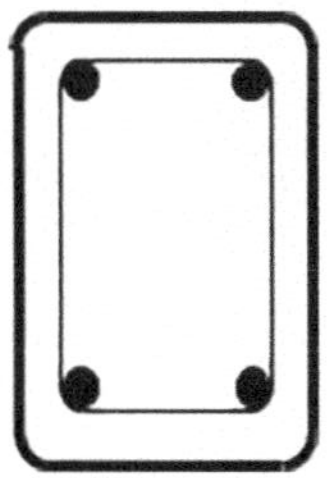	$f_{fwd} = 0.8 k_r f_{fud}$ (4.3) Where k_r is calculated using Equation 4.4 (*BS EN1992-1-1 2023*, Formula J.13) f_{fud} is calculated using Equation 2.1 (*BS EN 1992-1-1 2023*, Formula J.2)	The factor 0.8 accounts for long-term loading on the CFRP system. k_r is a reduction factor, which accounts for strength reduction in wrapping a CFRP system around corners and also takes into account stress concentrations where the CFRP strengthening crosses a shear crack. $k_r = \begin{cases} \frac{r_c}{50}\left(2 - \frac{r_c}{50}\right), r_c < 50\text{mm} \\ 1.0, r_c \geq 50\text{mm} \end{cases}$ (4.4)
Open discrete CFRP system 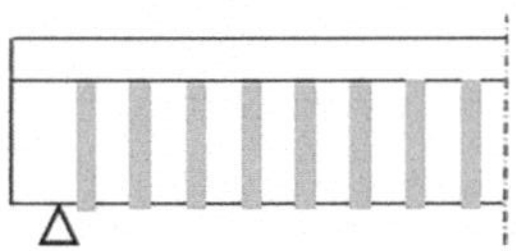	There are two conditions for determining f_{fwd} for open discrete systems, which depend on the anchorage that can be provided. They are (a) where $l_{bf} \leq l_{bf,max,k}$ for **all** CFRP strips $f_{fwd} = \frac{2}{3} \frac{n s_f}{l_{bf,max,k}\left[(\cot\theta + \cot\alpha_f)\sin\alpha_f\right]} f_{bfRd}$ (4.5) (b) where $l_{bf} \leq l_{bf,max,k}$ for **some** CFRP strips $f_{fwd} = \left[1 - \left(1 - \frac{2}{3} \frac{m s_f}{l_{bf,max,k}\left[(\cot\theta + \cot\alpha_f)\sin\alpha_f\right]}\right)\frac{m}{n}\right] f_{bfRd}$ (4.6) f_{bfRd} is calculated in accordance with *BS EN 1992-1-1:2023*, J.11.1.1.3	As illustrated in Figure 4.2, m is the number of CFRP stirrups crossing a crack with $l_{bf} \leq l_{bf,max,k}$, defined in Equation 4.7 $m = integer\left(\frac{l_{bf,max,k}\left[(\cot\theta + \cot\alpha_f)\sin\alpha_f\right]}{s_f}\right)$ (4.7) Also as illustrated in Figure 4.2, n is the number of all CFRP stirrups crossing a crack defined in Equation 4.8 $n = integer\left(\frac{h_f(\cot\theta + \cot\alpha_f)}{s_f}\right)$ (4.8)

Table 4.1 Continued

Type of System	Design Shear Resistance of Strengthening System (f_{fwd})	Comments
Open continuous CFRP system 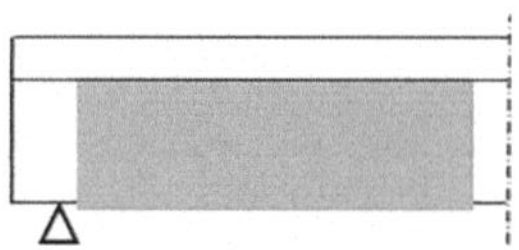	For open continuous sheets, two provisions are again required depending on the anchorage that can be provided. They are both summarised in Equation 4.9. $$f_{fwd} = \begin{cases} \frac{2}{3}\frac{\frac{h_f}{sin\alpha_f}}{l_{bf,max,k}} f_{bfRd} \quad for \quad \frac{h_f}{sin\alpha_f} < l_{bf,max,k} \\ \left(1 - \frac{1}{3}\frac{l_{bf,max,k}}{\frac{h_f}{sin\alpha_f}}\right) f_{bfRd} \quad for \quad \frac{h_f}{sin\alpha_f} \geq l_{bf,max,k} \end{cases} \quad (4.9)$$ f_{bfRd} is calculated in accordance with *BS EN 1992-1-1:2023*, J.11.1.1.3	An open continuous strengthening system may have only part of the system fully anchored. This is reflected in the two provisions for f_{fwd}.

Figure 4.2 Terminology for shear strengthening with EBR CFRP strips (Reproduced and modified from *BS EN 1992-1-1:2023*)

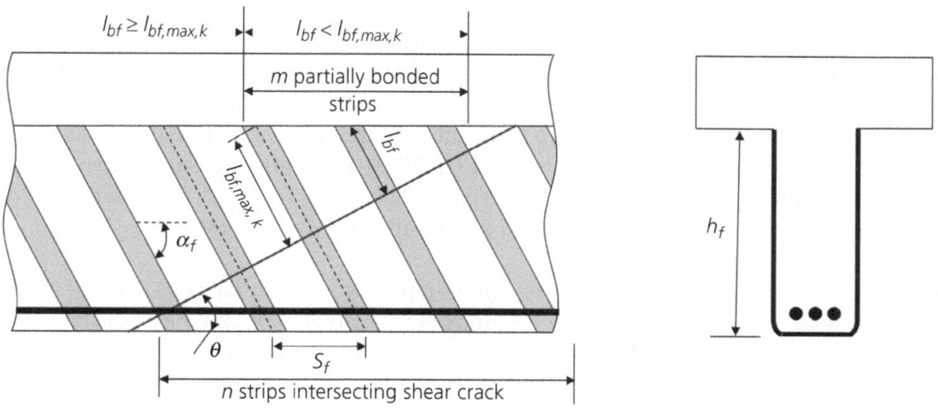

Note – the fibre orientation of the strengthening system in this figure is parallel to the edge of the illustrated strips

Note that Equation 4.2 does remove a double-counted term between Formulae J.16 and J.17 that accounts for the angle of the fibres relative to the longitudinal axis of the member in *BS EN 1992-1-1:2023*. It is understood that the UK Published Document to *BS EN 1992-1-1:2023* will clarify this.

Design shear strength of strengthening system

f_{fwd} in Equation 4.2 is the design shear strength of the strengthening system. Depending on the strengthening system deployed, f_{fwd} can depend on either the design tensile strength of the CFRP system, f_{fud}, or the design bond resistance, f_{bfRd}. In instances where open strengthening systems are used, it is likely that at least part of the shear strengthening system will not have achieved full bond length, and the bond resistance will govern the shear resistance. Table 4.1 presents the values of f_{fwd} for each type of shear strengthening system that can be used in accordance with this standard, which takes into consideration the potential for reduced strength resulting from anchorage lengths which are less than $l_{bf,max,k}$ where appropriate.

Where open systems are used, the bond strength between the CFRP strengthening system and concrete substrate will limit the design shear strength. There are two different scenarios considered in the standard to cater for this, as follows.

- Discrete shear strengthening systems – this will generally be implemented by externally bonded CFRP strips in accordance with this standard. In this instance, strips may either have full anchorage length, $l_{bf,max,k}$, in the compression chord above a shear crack, or not. As shown in Table 4.1, the provisions allow an estimation of the number of strips, n, that cross a crack and the number of strips, m, with anchorage length, l_{bf}, into the compression zone above a shear crack having length less than the characteristic minimum anchorage, $l_{bf,max,k}$, needed to generate the full bond resistance. *BS EN 1992-1-1:2023*, Formulae J.21 and J.22 may be used to determine values of m and n.

The provisions also allow for shear strengthening strips to be orientated at an angle α_f from the longitudinal axis of the member being strengthened. $\alpha_f = 90°$ when the strengthening is applied perpendicular to the longitudinal axis of the member.

- Continuous shear strengthening systems – in the case of continuous shear strengthening systems, similar to discrete systems, part of the sheet used for the strengthening system will likely be applied where there is theoretically insufficient anchorage into the compression region beyond a shear crack to achieve the full bond strength. Limitation of the bond is taken into account by considering the length of the CFRP strengthening measured perpendicular to the longitudinal axis, and two cases are given, as shown in Table 4.1. Orientation of α_f for continuous systems (sheets) is identical to that for discreet systems (strips).

Fully wrapping a member with CFRP significantly improves shear resistance. The end anchorage failure mechanism described in Chapter 3 cannot occur. The reduction in CFRP resistance for end anchorage failure does not need to be considered in this case, and a direct relationship between f_{fwd} and f_{fud} in Equation 4.3 (*BS EN 1992-1-1:2023*, Formula J.18) is used to reflect this. Factors are applied to the design shear resistance of the strengthening system including a long-term strength reduction factor (0.8) and a factor for strength reduction when wrapping the CFRP around corners that also accounts for stress concentrations at the intersection between the CFRP strengthening system and shear cracks, k_r.

Effective strut angle

BS EN 1992-1-1:2023, J.8.2.3(4) requires that a strut angle of 45° is applied for determining τ_{Rd} and $\tau_{Rd,f}$. It indicates that more rigorous methods of analysis can be applied for determining this angle. This provision is though consistent with the approach in ACI 318-19 (ACI, 2019). It is different from the approach taken in Australian Standard AS 5100.8:2017 (AS, 2017), where the strut angle calculated for assessment of the shear capacity is taken. AS5100.8 does however use different methods for determining the strut angle.

TR55 (Concrete Society, 2013) does permit shallower strut angles for the unstrengthened section to be superimposed with a CFRP resistance based on a strut angle of 45°. The provisions effectively take a weighted average of the strut angle for the strengthened and unstrengthened member by superimposing pre- and post-strengthened resistance. This does, however, require that the shear steel should yield prior to failure of the CFRP (including bond failure). If a designer does consider this method, they should consult TR55 (Concrete Society, 2013) for specific provisions that apply.

Additional forces in the longitudinal reinforcement

As stated in *BS EN 1992-1-1:2023*, J.3(1), all clauses in Sections 1 to 14 apply unless specifically omitted or supplemented by Annex J. Therefore, Clause 8.2.3(8) of the main standard applies in determining the additional tensile force in longitudinal reinforcement resulting from the presence of an applied shear where required for design. The value of V_{Ed} used in the expressions in this section of the code will be the shear action effect at ULS when strengthened.

The application of the shift rule in *BS EN 1992-1-1:2023*, 12.3.2 negates the need to consider the additional force in the CFRP in the anchorage zone.

Spacing of CFRP strips

BS EN 1992-1-1:2023 has no specific provisions on the spacing of CFRP strips applied for resisting shear. As stated in TR55 (Concrete Society, 2013), CFRP strips applied for resisting shear will need to be positioned so as to ensure they intersect shear cracks and that shear cracks do not form between strips. This is further discussed in Chapter 9 of this book.

Compatibility and deformation capacity

Underpinning the code relationships is strain compatibility between the CFRP strengthening system and steel internal reinforcement. fib Bulletin No. 90 (fib, 2019), which aligns closely with the provisions of *BS EN 1992-1-1:2023*, indicates that an assumption is inherent in the provisions that there is sufficient ductility in the steel shear reinforcement for the CFRP to reach its design strength f_{fwd} prior to rupture of the shear stirrups. Ensuring this is the case will mean that the steel shear reinforcement will not rupture before the CFRP reaches its rupture strength, also validating the superposition applied in *BS EN 1992-1-1:2023*.

Equation 4.10 provides a simple mathematical condition that may be used to validate this condition. This equation conservatively assumes that the strain in the steel shear reinforcement at strengthening is very small compared with the ultimate strain ε_{uk}. It also uses the characteristic ultimate strain of the CFRP rather than the smaller design value of strain, which results in a more onerous check.

$$\frac{f_{fuk}}{E_f} < \varepsilon_{uk} \tag{4.10}$$

Where

ε_{uk} is the characteristic maximum elongation of the steel stirrups at the maximum load and can be taken as shown in Table 4.2 for the ductility classes used in *BS EN 1992-1-1:2023*.

Additionally, the superposition principle effectively implies that the reinforcement will yield ahead of debonding failure of the CFRP. It is noted though that the strain at CFRP debonding will likely be less than the yield strain of reinforcement. This suggests that shear strengthening should be applied in instances where the stirrups are highly utilised under existing loading conditions. It is noted that the superposition principle has been widely adopted in various international guidance documents, and is supported by substantial experimental database. It does though reinforce the need to be cautious in seeking to reduce the strut angle used in design.

Table 4.2 Characteristic maximum elongation for maximum load for *BS EN 1992-1-1:2023* ductility classes (Adapted from *BS EN 1992-1-1:2023*)

	Ductility Class A	Ductility Class B	Ductility Class C
Characteristic maximum elongation at the maximum load, ε_{uk}	2.5 %	5 %	7.5 %

Example EBR Shear 1: Shear strengthening with EBR CFRP strips

Determine the maximum design shear resistance of the beam with strengthening arrangement shown in Figure 4.3, assuming a persistent design situation and the following parameters

$$V_{Ed} = 385 \text{ kN}$$

$$f_{ck} = 30 \text{ MPa}$$

$$f_{cm} = 38 \text{ MPa}$$

$$f_{ctm,surf} = 2.0 \text{ MPa}$$

$$f_{yk} = 400 \text{ MPa}$$

$$f_{fuk} = 2900 \text{ MPa}$$

$$E_f = 170\,000 \text{ MPa}$$

$$k_{sys,b1} = k_{sys,b2} = k_{sys,b3} = 1$$

Cover to stirrups is 40 mm (top and bottom).

In the unstrengthened case, θ is limited by conditions of ductility (*BS EN 1992-1-1:2023*, 8.2.3(4)) and availability of longitudinal reinforcement (*BS EN 1992-1-1:2023*, 8.2.3(8)), such that the assessed value of unstrengthened shear resistance is calculated using $\theta = 36°$.

Use the recommended values from *BS EN 1992-1-1:2023*, 5.1.6 for $k_{tc} = 0.85$ and $k_{tt} = 0.7$ to determine the design concrete compressive strength.

$$\gamma_c = 1.5$$

First, calculate CFRP anchorage properties, which are needed in the determination of f_{fwd}.

Figure 4.3 Beam details for EBR shear strengthening Example 1

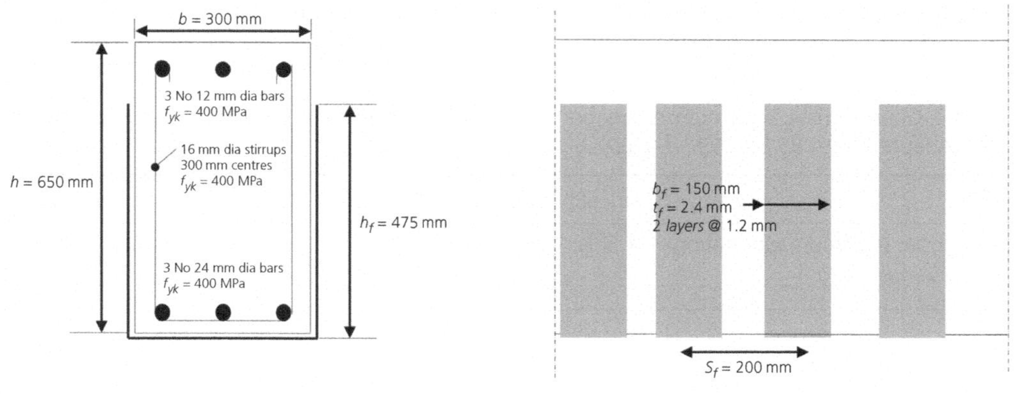

According to Equation 3.4 (Formula J.39 in *BS EN 1992-1-1:2023*), the length at which the maximum design bond resistance is reached is

$$l_{bf,max,k} = \frac{2}{k_{sys,b3}}\sqrt{\frac{E_f\, t_f\, s_{f0k}}{\tau_{f1k}}}$$

As set out in the assumptions in the example and Figure 4.3

$E_f = 170\,000$ MPa, $t_f = 2.4$ mm, $k_{sys,b3} = 1.0$

Note, no reduction to t_f is required in accordance with *BS EN 1992-1-1:2023*, J.5.2(4).

From Equation 3.6 (*BS EN 1992-1-1:2023*, Formula J.41)

$$s_{f0k} = 0.2 k_{sys,b2}$$

As set out in the assumptions in the example

$$k_{sys,b2} = 1.0$$

Therefore

$$s_{f0k} = 0.2$$

From Equation 3.5 *BS EN 1992-1-1:2023*, Formula J.42

$$\tau_{f1k} = 0.37 k_{sys,b2}\sqrt{f_{cm} f_{ctm,surf}}$$

As set out in the assumptions in the example

$f_{cm} = 38$ MPa, $f_{ctm,surf} = 2.0$ MPa, $k_{sys,b2} = 1.0$

Therefore

$$\tau_{f1k} = 0.37 \times 1.0 \times \sqrt{38 \times 2.0} = 3.2 \text{ MPa}$$

Therefore

$$l_{bf,max,k} = \frac{2}{1.0}\sqrt{\frac{170\,000 \times 2.4 \times 0.2}{3.2}} = 319 \text{ mm}$$

According to Equation 3.7 and Equation 3.8 (Formulae J.38 and J.40 in *BS EN 1992-1-1:2023*), the design bond strength of the anchorage, where full anchorage is provided, is

$$f_{bfRd} = \frac{\sqrt{\eta_{cc} k_{tc} k_{tt}}}{\gamma_{BA}}\sqrt{\frac{E_f\, \tau_{f1k}\, s_{f0k}}{t_f}}$$

According to Formula 5.4 in *BS EN 1992-1-1:2023*

$$\eta_{cc} = \left(\frac{f_{ck,ref}}{f_{ck}}\right)^{\frac{1}{3}} \leq 1.0$$

Where

$f_{ck,ref} = 40\,\text{MPa}$ (*BS EN 1992-1-1:2023*, 5.1.6) and $f_{ck} = 30$ MPa, as stated in the assumptions of the example.

Therefore

$$\eta_{cc} = \left(\frac{40}{30}\right)^{\frac{1}{3}} > 1.0$$

Therefore

$$\eta_{cc} = 1.0$$

Therefore

$$f_{bfRd} = \frac{\sqrt{1.0 \times 0.85 \times 0.7}}{1.5}\sqrt{\frac{170\,000 \times 3.2 \times 0.2}{2.4}} = 110\ \text{MPa}$$

To use the provisions of *BS EN 1992-1-1:2023*, Annex J, a strut angle of $\theta = 45°$ should be assumed. The unstrengthened strut angle $\theta = 36°$ should not be used, and a weighted average of strut angles should not be used unless it can be verified that the links will yield prior to CFRP failure.

Equation 4.1 (*BS EN 1992-1-1:2023*, Formula J.15) gives

$$\tau_{Rd,CFRP} = \tau_{Rd} + \tau_{Rd,f}$$

The sectional shear stress in accordance with *BS EN 1992-1-1:2023*, Formula 8.18 is

$$\tau_{Ed} = \frac{V_{Ed}}{b_w z}$$

Where

$$z = 0.9d = 0.9 \times (650 - 40 - 16 - 12) = 524\ \text{mm}$$

Therefore

$$\tau_{Ed} = \frac{385\,000}{300 \times 524} = 2.45\ \text{MPa}$$

Apply *BS EN 1992-1-1:2023*, Formula 8.42

$$\tau_{Rd,sy} = \rho_w f_{ywd}\, cot\theta$$

$$\theta = 45°$$

$$f_{ywd} = \frac{f_{ywk}}{\gamma_s} = \frac{400}{1.15} = 348 \text{ MPa}$$

$$\rho_w = \frac{A_{sw}}{b_w s} = \frac{402}{300 \times 300} = 0.0045$$

$$\tau_{Rd,sy} = \rho_w f_{ywd} cot\theta = 1.55 \text{ MPa}$$

Checking that the compression strut will not crush, *BS EN 1992-1-1:2023*, Formula 8.44 is applied

$$\sigma_{cd} = \tau_{Ed} \left(\cot\theta + \tan\theta\right) \leq \nu f_{cd}$$

f_{cd} is determined using *BS EN 1992-1-1:2023*, Formulae 5.3 and 5.4

$$f_{cd} = \eta_{cc} k_{tc} \frac{f_{ck}}{\gamma_c}$$

Therefore

$$f_{cd} = k_{tc} \frac{f_{ck}}{\gamma_c} = 0.85 \times \frac{30}{1.5} = 17 \text{ MPa}$$

According to *BS EN 1992-1-1:2023*, 8.2.3(6) $\nu = 0.5$

Therefore

$$\nu f_{cd} = 0.5 \times 17 = 8.5 \text{ MPa} > \tau_{Ed} = 2.45 \text{ MPa}$$

Therefore, there is no crushing of the shear compression strut.

At $\theta = 45°$

$$\tau_{Rd} = \tau_{Rd,sy} = 1.55 \text{ MPa}$$

Combining *BS EN 1992-1-1:2023*, Formulae J.16 and J.17 with $\alpha_f = 90°$ and $\theta = 45°$ for CFRP strips, Equation 4.2 becomes

$$\tau_{Rd,f} = \frac{2t_f b_f}{s_f} \frac{f_{fwd}}{b_w}$$

Check if some or all strips have an anchorage length into the compression zone $\geq l_{bf,max,k}$.

Consider the terms *m* and *n* in *BS EN 1992-1-1:2023*, Formulae J.22 and J.21, respectively, and acknowledge the fact that $\theta = 45°$ and $\alpha_f = 90°$.

The number of strips crossing the critical shear crack (Equation 4.8)

$$n = integer\left(\frac{h_f}{s_f}\right) = integer\left(\frac{475}{200}\right) = 2$$

The number of strips that have anchorage length $\geq l_{bf,max,k}$ into the compression zone (Equation 4.7)

$$m = integer\left(\frac{l_{bf,max,k}}{s_f}\right) = integer\left(\frac{319}{200}\right) = 1$$

Therefore "some" strips intersect the crack and Equation 4.6 applies for determining f_{fwd}.

Simplifying *BS EN 1992-1-1:2023*, Formula J.20 (Equation 4.6) with $\alpha_f = 90°$ and $\theta = 45°$, the following is obtained

$$f_{fwd} = \left[1 - \left(1 - \frac{2}{3}\frac{ms_f}{l_{bf,max,k}}\right)\frac{m}{n}\right] f_{bfRd}$$

$l_{bf,max,k} = 319$ mm and $f_{bfRd} = 110$ MPa

Therefore

$$f_{fwd} = \left[1 - \left(1 - \frac{2}{3} \times \frac{1 \times 200}{319}\right) \times \frac{1}{2}\right] \times 110 = 78 \text{ MPa}$$

and

$$\tau_{Rd,f} = \frac{2 \times 2.4 \times 150}{200} \times \frac{78}{300} = 0.94 \text{ MPa}$$

which leads to

$$\tau_{Rd,CFRP} = 0.94 + 1.55 = 2.49 \text{ MPa} > \tau_{Ed} = 2.45 \text{ MPa}$$

Therefore, the strengthening system is adequate.

Example EBR Shear 2: Shear strengthening with open EBR CF Sheets

Determine the maximum design shear resistance that the beam in Example EBR Shear 1 could take if it was strengthened as shown in Figure 4.4, instead of using CFRP strips. For the CF sheets, assume that

$f_{fuk} = 2400$ MPa

$E_f = 225\,000$ MPa

$l_{bf,max,k} = 192$ mm

$f_{bfRd} = 241$ MPa

The cross-sectional area of the fibres per metre of a single layer of CF sheet given by the supplier is 660 mm²/m. Therefore, for a metre width of a CF sheet

$A_{f,1} = 660$ mm^2 per single layer

Equation 4.1 (*BS EN 1992-1-1:2023*, Formula J.15) gives

$$\tau_{Rd,CFRP} = \tau_{Rd} + \tau_{Rd,f}$$

Figure 4.4 Beam details for EBR shear strengthening Example 2

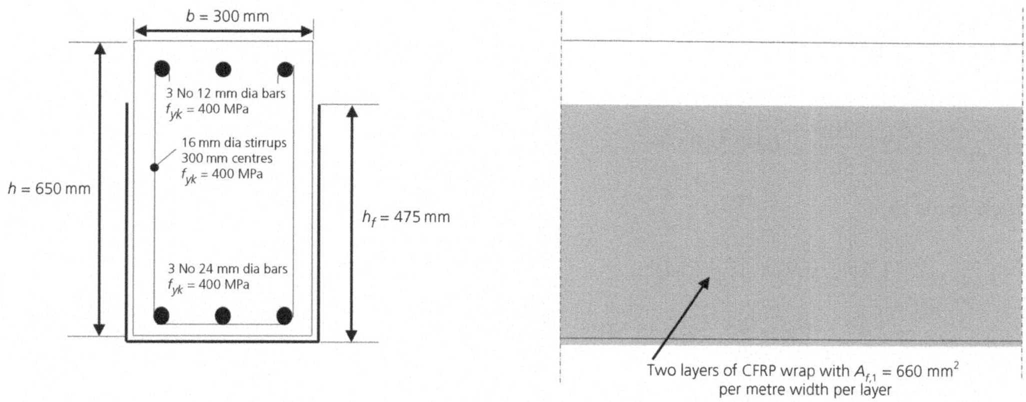

τ_{Rd} for the unstrengthened section remains unchanged from Example EBR Shear 1, i.e.

$$\tau_{Rd} = 1.55 \text{ MPa}$$

Combining *BS EN 1992-1-1:2023*, Formulae J.16 and J.17 with $\alpha_f = 90°$ and $\theta = 45°$ for continuous CF sheets, the following is obtained

$$\tau_{Rd,f} = 2t_f \frac{f_{fwd}}{b_w}$$

As per *BS EN 1992-1-1:2023*, J.5.2(4), where several layers of sheets are applied, the effective thickness is obtained from

$$t_f = n_f^{\ k_f} \frac{A_f}{b_f}$$

Where

n_f is the number of sheets, $n_f = 2$

k_f is an exponent depending on the number of sheets, for this example with 2 sheets $k_f = 1$

$$t_f = n_f^{\ k_f} \frac{A_{f,1}}{b_f} = 2^1 \frac{660}{1000} = 1.32 \text{ mm}$$

No reduction in effective thickness or area applies.

As $h_f > l_{bf,max,k}$, the following applies when adopting $\alpha_f = 90°$ in *BS EN 1992-1-1:2023*, Formula J.23 (Equation 4.9)

$$f_{fwd} = \left(1 - \frac{1}{3}\frac{l_{bf,max,k}}{\frac{h_f}{sin\alpha_f}}\right) f_{bfRd}$$

which yields

$$f_{fwd} = \left(1 - \frac{1}{3} \times \frac{\frac{192}{475}}{1}\right) \times 241 = 209 \text{ MPa}$$

$$\tau_{Rd,f} = 1.32 \times \frac{209}{300} = 0.92 \text{ MPa}$$

which leads to

$$\tau_{Rd,CFRP} = 1.55 + 0.92 = 2.47 \text{ MPa}$$

$$\tau_{Rd,CFRP} = 2.47 \text{ MPa} > \tau_{Ed} = 2.45 \text{ MPa}$$

Therefore, the strengthening system is ok.

Example EBR Shear 3: Shear strengthening with closed EBR CF sheets

Determine the maximum design shear resistance that the beam in Example EBR Shear 1 could take if it was strengthened as shown in Figure 4.5, instead of using CFRP strips. For the CF sheets, assume that

$$f_{fuk} = 3600 \text{ MPa}$$

$$E_f = 230\,000 \text{ MPa}$$

$$r_c = 25 \text{ mm}$$

In this case, the design shear strength of the strengthening system is calculated using *BS EN 1992-1-1:2023*, Formula J.18 (Equation 4.3) as

$$f_{fwd} = 0.8 k_r f_{fud}$$

Figure 4.5 Beam details for EBR shear strengthening Example 3

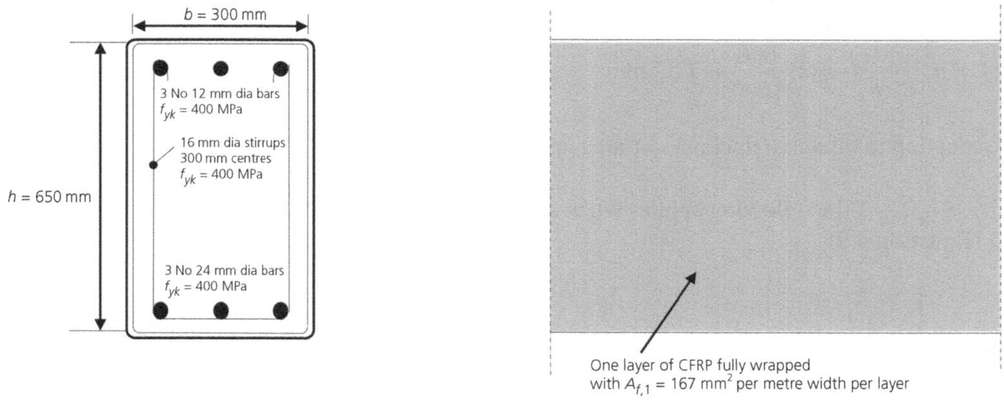

BS EN 1992-1-1:2023, Formula J.13 (Equation 4.4) is used to calculate k_r, which for $r_c = 25$ mm, as follows

$$k_r = \frac{r_c}{50}\left(2 - \frac{r_c}{50}\right) = 0.75$$

f_{fud} is determined in accordance with *BS EN 1992-1-1:2023*, J.5.3(1) – that is,

$$f_{fud} = \eta_f \frac{f_{fuk}}{\gamma_f}$$

Where

$\eta_f = 0.7$

$\gamma_f = 1.4$ for CF sheets

Therefore

$$f_{fud} = 0.7 \times \frac{3600}{1.4} = 1800 \text{ MPa}$$

Therefore

$$f_{fwd} = 0.8 \times 0.75 \times 1800 = 1080 \text{ MPa}$$

With all other parameters being equal to those in Example EBR Shear 2

$$\tau_{Rd,f} = 2 \times 0.167 \times \frac{1080}{300} = 1.20 \text{ MPa}$$

which leads to

$$\tau_{Rd,CFRP} = 2.75 \text{ MPa} > \tau_{Ed} = 2.45 \text{ MPa}$$

Torsion, combined actions and punching shear

The code does not permit inclusion of the beneficial effects of CFRP strengthening in determining torsional resistance. It also requires that resistances for use in interaction of combined actions do not include the benefit of CFRP, nor inclusion of CFRP in determining punching shear resistance. There simply was not an exhaustive enough database to allow robust code provisions to be developed for this application. Other guidance such as TR55 (Concrete Society, 2013) or fib Bulletin No. 90 (fib, 2019) may be consulted if such strengthening is to be undertaken.

Kaloyana Kostova and Craig Giaccio
ISBN 978-1-83549-935-1
https://doi.org/10.1108/978-1-83549-934-420251005

Chapter 5
Flexural design

Introduction

The basic provisions of *BS EN 1992-1-1:2023* (BSI, 2023b), 8.1.1 apply to reinforced or prestressed concrete members strengthened with CFRP systems. *BS EN 1992-1-1:2023*, Annex J gives further provisions for including the CFRP in calculating bending resistance of a strengthened cross section, which are as follows.

- The compressive strength of the CFRP should not be considered in determining flexural strength.
- The slip at the interface between the CFRP strengthening system and concrete can be ignored.

This chapter provides methods for determining the flexural resistance of a member strengthened with CFRP. It outlines the following.

- Determining sectional resistance of a member.
- Additional provisions for the bond of EBR systems that need to be considered when determining the flexural resistance of a member strengthened with CFRP, including
  - end anchorage
  - intermediate crack debonding
  - end cover separation
  - shear induced crack separation.
- Additional provisions for the bond of NSM systems that need to be considered when determining the flexural resistance of a member strengthened with CFRP, including
  - end anchorage
  - end cover separation
  - shear induced crack separation.
- A brief discussion on *BS EN 1992-1-1:2023* requirements regarding moment redistribution.

Ultimate cross-sectional resistance

When determining the resistance of a concrete member strengthened with CFRP, the following strain limitations shall not be exceeded when determining the design resistance to bending moment.

- The ultimate compressive strain of concrete, ε_{cu}, as defined in *BS EN 1992-1-1:2023*, 8.1.2.
- The ultimate strain of reinforcing steel, pre-stressing or post-tensioning, ε_{ud}.
- The strain at the design tensile strength of the CFRP, ε_{fud}, determined using Equation 2.2.
- The maximum strain corresponding to an intermediate crack debonding failure calculated in accordance with *BS EN 1992-1-1:2023*, J.11.1.2, as outlined in Chapter 3, which is only required for members strengthened in flexure with EBR CFRP systems.

When checking strain limits, changes in sectional properties that occur as a result of the application of the CFRP need to be considered. As illustrated in Figure 5.1, this can be done using the principle of superposition of strains (and stresses).

A strain compatibility-based method will be the most reliable method to calculate sectional bending resistance of a member strengthened with CFRP. A flow chart showing the steps to determine the flexural resistance for a member using an iterative strain-based method with CFRP is shown in Annex A.

The simplified ultimate limit state (ULS) method with rectangular stress block may be used for cross-sectional analysis as stated in *BS EN 1992-1-1:2023*, 8.1.2. CFRP also can be transformed into an equivalent steel area to determine flexural resistance using these provisions.

With three materials rather than two, it may be beneficial to ensure the neutral axis is correctly located. fib Bulletin No. 90 (fib, 2019) provides a method that may be used to determine the resultant compression force and lever arm using the parabolic relationship in *BS EN 1992-1-1:2023*, 8.1.2 without the need for detailed iterative calculations. Equations used in the method are directly derived from the stress-strain relationship presented in Section 8 of *BS EN 1992-1-1:2023* for the design of cross sections, therefore they have a sound theoretical basis for use. These provisions are used for comparisons in some of the examples that follow. A brief explanation of the method follows. The full derivation of the provisions is provided in Annex D. They are suited only to sections that can be represented with a constant width of the compression region at ULS. The parabolic distribution is used in the examples of this book to illustrate that it is quite simple to use.

Consider first the general equilibrium condition of a rectangular cross section shown in Figure 5.2.

Forces shown in Figure 5.2 can be summed as shown in Equation 5.1.

$$F_c + \sigma_{sc.1} A_{sc.1} = \sigma_{st.1} A_{st.1} + \sigma_f A_f \tag{5.1}$$

Figure 5.1 Illustration of superposition required to determine strain in cross section of member subjected to flexure

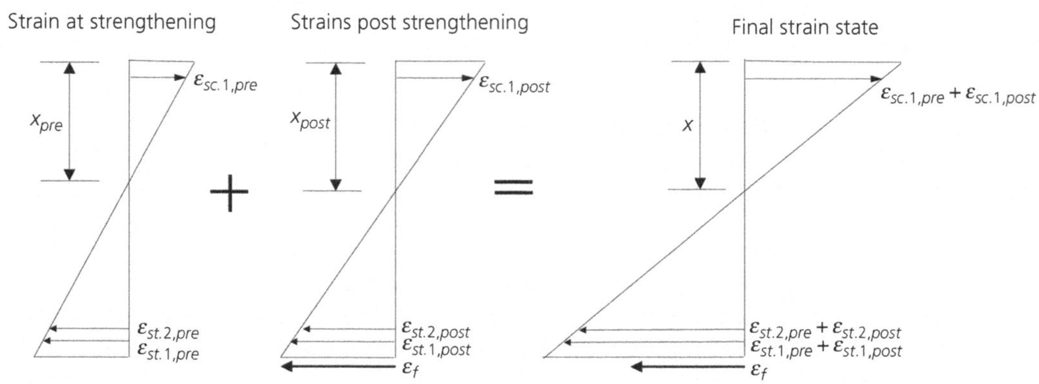

Figure 5.2 Force balance in a doubly reinforced cross section strengthened in flexure with EBR CFRP (Adapted from fib Bulletin No. 90 (fib, 2019))

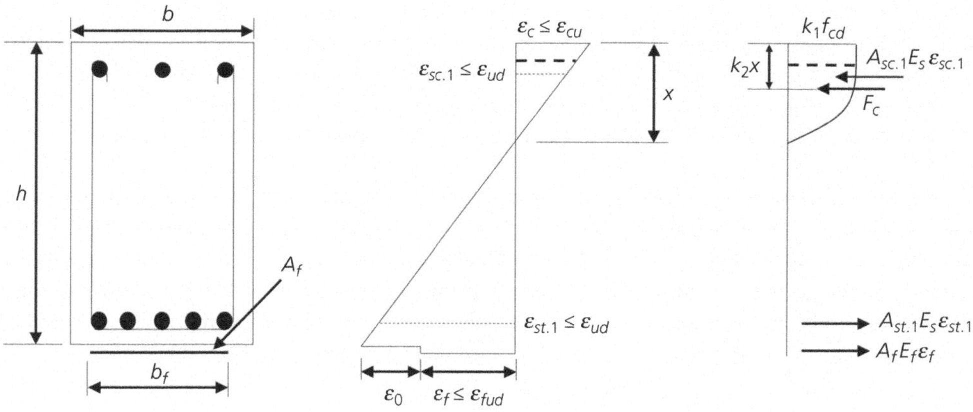

- - - - Denotes location in cross section at which concrete compressive strain $\varepsilon = \varepsilon_{c2} = 0.002$

Where, as shown in Equation 5.2

$$F_c = k_1 f_{cd} bx$$

Where

$$k_1 = \begin{cases} 1000\varepsilon_c \left[0.5 - \dfrac{250\varepsilon_c}{3}\right], & \varepsilon_c < 0.002 \\ 1 - \dfrac{1}{1500\varepsilon_c}, & 0.002 \leq \varepsilon_c \leq 0.0035 \end{cases} \tag{5.2}$$

The accompanying internal moment for an EBR CFRP system, associated with a given force balance, can then be obtained by summing moments about the position of the compression force, as shown in Equation 5.3.

$$\mathrm{M} = \sigma_{st.1} A_{st.1} \left(d_{st.1} - k_2 x\right) + E_f A_f \varepsilon_f \left(h - k_2 x\right) + \sigma_{sc.1} A_{sc.1} \left(k_2 x - d_{sc.1}\right) \tag{5.3}$$

Where, as shown in Equation 5.4

$$k_2 = \begin{cases} \dfrac{2 - 250\varepsilon_c}{6 - 1000\varepsilon_c}, & \varepsilon_c < 0.002 \\ \dfrac{1000\varepsilon_c \left(3000\varepsilon_c - 4\right) + 2}{2000\varepsilon_c \left(3000\varepsilon_c - 2\right)}, & 0.002 \leq \varepsilon_c \leq 0.0035 \end{cases} \tag{5.4}$$

To find the flexural resistance, one simply needs to determine the strain distribution that produces the highest internal moment within the strain limits that apply to the constituent materials. To achieve this, the limits of strain associated with CRP rupture, ε_{fud}, concrete crushing, ε_{cu}, and the design strain of the reinforcement of prestressing at the maximum load, ε_{ud}, need to be evaluated. Each of these limiting strains may be used as a condition to assess resistance by using an iterative method to achieve a force balance using Equation 5.1. The corresponding internal moment determined using Equation 5.3 is the resistance associated with this strain distribution. The moment resistance, M_{Rd}, is the maximum value of the internal moment where only one of the three governing strain conditions is reached.

While omitting the strain superposition shown in Figure 5.2 may be conservative in some instances, it may introduce inefficiency into the design if there is a relatively high proportion of the ULS load present at the time of strengthening. fib Bulletin No. 90 (fib, 2019) suggests that strain superposition need not be undertaken if the moment at the section at strengthening is less than the cracking moment, M_{cr}, a decision that can be made at the designer's discretion. The examples in this book consider the pre-strain to ensure that an accurate representation of the principles is presented.

Considerations for EBR CFRP systems

The resistance of a member subjected to bending and strengthened with EBR CFRP will often be limited by failure of the bond between the CFRP and concrete strata. If the bond limits the capacity, this will mean that the CFRP will not be able to reach its full design tensile strength, nor reach the ultimate cross-sectional flexural resistance. The checks which need to be undertaken to assess if the bond is limiting are described in Chapter 3 of this book, and are listed below.

- End anchorage resistance (*BS EN 1992-1-1:2023*, J.11.1.1.1) including relevant curtailment requirements (*BS EN 1992-1-1:2023*, J.11.1.2.2).
- Intermediate crack debonding resistance (*BS EN 1992-1-1:2023*, J.11.1.2.3).
- End cover separation (*BS EN 1992-1-1:2023*, J.11.1.2.4).
- Shear induced separation (*BS EN 1992-1-1:2023*, J.11.1.2.5).

Bond failure will often occur at a section some distance from the point of maximum moment, meaning that intermediate crack debonding checks should be undertaken between each crack (spaced at $s_{cr,min}$) within the strengthened area. These checks also require consideration of pre-strain and need a strain compatibility approach to achieve an efficient design. Undertaking these checks with strain compatibility-based iterative procedures is computationally intensive, which is best undertaken using a bespoke computer program. A flow chart of steps that need to be undertaken to complete this check is provided in Annex A.

A simplified method is presented in this chapter, which is better suited to hand calculation. This chapter also provides commentary on how one can identify where intermediate crack debonding might govern the member flexural resistance.

Simplified method for determining intermediate crack debonding action effects

The approach here is an extension of the method provided in fib Bulletin No. 90 (fib, 2019). By transforming the CFRP to an equivalent steel area, the equivalent total area of reinforcement can be found as shown in Equation 5.5.

$$A_{equiv} = \frac{E_f}{E_s} A_f + \sum_{i=1}^{n} A_{st.i} \tag{5.5}$$

Where

n is the number of layers of reinforcement in the tension region of a member in flexure.

The equivalent steel area can further be simplified into a single layer. This is achieved by using a weighted averaging procedure to determine an equivalent effective depth of the section as shown in Equation 5.6.

$$d_{equiv} = \frac{d_f E_f A_f + \sum_{i=1}^{n} d_{st.i} E_s A_{st.i}}{E_f A_f + \sum_{i=1}^{n} E_s A_{st.i}} \tag{5.6}$$

To determine the inner lever arm between the tension and compression forces in a member subject to flexure, the fib Bulletin No. 90 (fib, 2019) method assumes that this lever arm is 80% of the equivalent effective depth, as shown in Equation 5.7, for debonding checks.

$$z_m = 0.8 d_{equiv} = 0.8 \frac{d_f E_f A_f + \sum_{i=1}^{n} d_{st.i} E_s A_{st.i}}{E_f A_f + \sum_{i=1}^{n} E_s A_{st.i}} \tag{5.7}$$

To determine the force in the reinforcement and the CFRP, tensile strains pre- and post-strengthening need to be evaluated, with consideration for the change in section properties that occur after strengthening.

The stress in the reinforcement prior to strengthening may be determined by using Equation 5.8. This is a common equation for determining the ULS capacity of a cross section. The inner lever arm of the section prior to strengthening is assumed to be 85% of the effective depth of the tension reinforcement.

$$\sigma_{s,pre} = \frac{M_{Ed,str}}{z A_{st}} = \frac{M_{Ed,str}}{0.85 d_{st} A_{st}} \leq f_{yd} \tag{5.8}$$

After strengthening, the total force in the combined tension components of the cross section, attributable to the additional bending moment applied after strengthening, may be calculated using Equation 5.9.

$$F_{equiv} = \frac{M_{post}}{z_m} = \frac{M_{Ed} - M_{Ed,str}}{z_m} \tag{5.9}$$

The force in the CFRP, or each layer of reinforcement, is simply calculated by reverse weighting the force out of the aggregated, transformed equivalent single layer of tension material with properties of the steel reinforcement.

The force in the CFRP at a given section can, therefore, be calculated as shown in Equation 5.10. This equation considers only a horizontal post elastic branch for the steel reinforcement.

$$F_{fEd} = \begin{cases} \dfrac{M_{Ed} - M_{Ed,str}}{z_m} \dfrac{d_f E_f A_f}{d_f E_f A_f + d_{st} E_s A_{st}}, & \sigma_{s,ULS} \leq f_{yd} \\ \dfrac{M_{Ed}}{z_m} - A_{st} f_{yd}, & \sigma_{s,ULS} > f_{yd} \end{cases} \quad (5.10)$$

The stress in the steel reinforcement at ULS can be determined by Equation 5.11.

$$\sigma_s = \sigma_{s,pre} + \frac{M_{Ed} - M_{Ed,str}}{z_m} \frac{d_{st} E_s}{d_f E_f A_f + d_{st} E_s A_{st}} \leq f_{yd} \quad (5.11)$$

This method ignores the contribution of any compression reinforcement. It can only be applied to under-reinforced sections – that is, those in which the steel reinforcement reaches yield before crushing of the concrete. It is only suitable for members that have a constant width of the compression region at ULS.

Analysis of the method undertaken by the authors in the preparation of this book suggests that the method generally underpredicts the inner lever arm, meaning that internal forces need to be greater to sustain an applied moment than they would be if calculated using a compatibility-based method. This means that force differences in CFRP are higher than they would be with an iterative method, in some cases considerably, meaning that the intermediate crack debonding checks are generally conservative when applying this method.

Case study – review of results from a detailed intermediate crack debonding check of a member strengthened in flexure with an EBR CFRP system

CFRP strengthening cannot be terminated until the stress in the CFRP is less than or equal to the design bond resistance, f_{bfRd}. Forces in the CFRP also need to consider a component from shear, which is implemented using the shift rule. This means that there can be CFRP reinforcement present in a strengthened member where the unstrengthened capacity, $M_{Rd,pre}$, is sufficient to resist applied moment, M_{Ed}. Figure 3.1 has been amended as shown in Figure 5.3 to reflect this.

Figure 5.3 Flexural strengthening zones for EBR CFRP systems

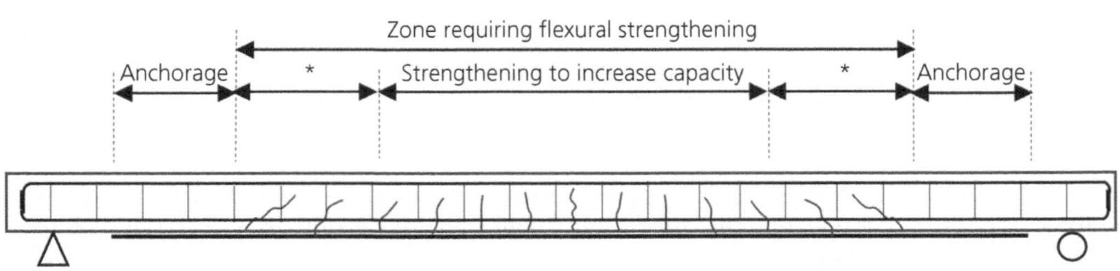

* Denotes that the CFRP will extend through this area until the design anchorage resistance can sustain forces generated from compatibility

Intermediate crack debonding checks require that the force difference between two cracks spaced at a distance $s_{cr,min}$ is less than the resistance as outlined in Chapter 3. The force difference in the EBR CFRP will be the greatest within the length of CFRP strengthening

- when moment gradients between cracks are highest under the load combination being considered
- where the reinforcement yields and a greater share of load is suddenly transferred to the CFRP. This effect will be greatest where a horizontal post-elastic branch is used in the strengthening design.

Figure 5.4 presents a case study that highlights where strengthening is needed for resistance, and where force differences in the CFRP between two sections spaced at $s_{cr,min}$ are highest. These diagrams have been extracted from Example Flexural Design B-1 in Annex B.

The first point to note is that the CFRP extends a significant distance beyond where it is needed to increase the flexural resistance of the member. This is because the termination point needs to be located so that the forces that the CFRP attracts from compatibility are less than the anchorage resistance, F_{bfRd}. For anchorage, the forces in the CFRP need to be adjusted to take into account the shift of the moment diagram, which is illustrated throughout the example.

The force in the CFRP picks up considerably at the point along the length of the member where the reinforcing steel yields, and cannot sustain additional load as the bending moment increases along the length. Note again that this example considers no strength gain in the reinforcement after it has yielded.

The figure shows that the significant increase in force difference occurs outside of the area where strengthening is required because of the application of the shift rule. In the example that corresponds to this case study, intermediate crack debonding does not govern the design, but it can. It is possible that a point outside of the area that needs strengthening is governed by intermediate crack debonding if the simplified method was used.

An approach to avoid undertaking intermediate crack debonding checks for all sections is to look at locations where the change in CFRP force is most significant. The first significant change in CFRP force will be at the section immediately before and after the point where the steel reinforcement yields (taking the shift term into account). Significant force differences in the CFRP force will also occur in the section where the reinforcement has yielded and where the moment gradient is at its highest along this length. This approach will not remove the fact that the intermediate crack debonding checks are computationally intensive, but illustrates that some engineering judgement could aid reducing the number of checks needed if an iterative strain compatibility-based solver is not available.

Considerations for NSM CFRP systems

Embedding rods or strips of CFRP into slots where the CFRP is fully encapsulated by adhesive beams, as is done for the NSM system, produces a bond resistance that is far greater than that of the EBR system. Because of this, failure mechanisms attributed to anchorage tend not to limit the resistance of members strengthened with an NSM CFRP system. *BS EN 1992-1-1:2023* does not require any debonding check for NSM systems.

The resistance of a member subjected to bending, strengthened with NSM CFRP will generally be limited by the sectional resistance. As is evident from *BS EN 1992-1-1:2023*, J.11.1.3(2) and (3),

Figure 5.4 Case study of a beam strengthened in flexure to identify areas where intermediate crack debonding governs

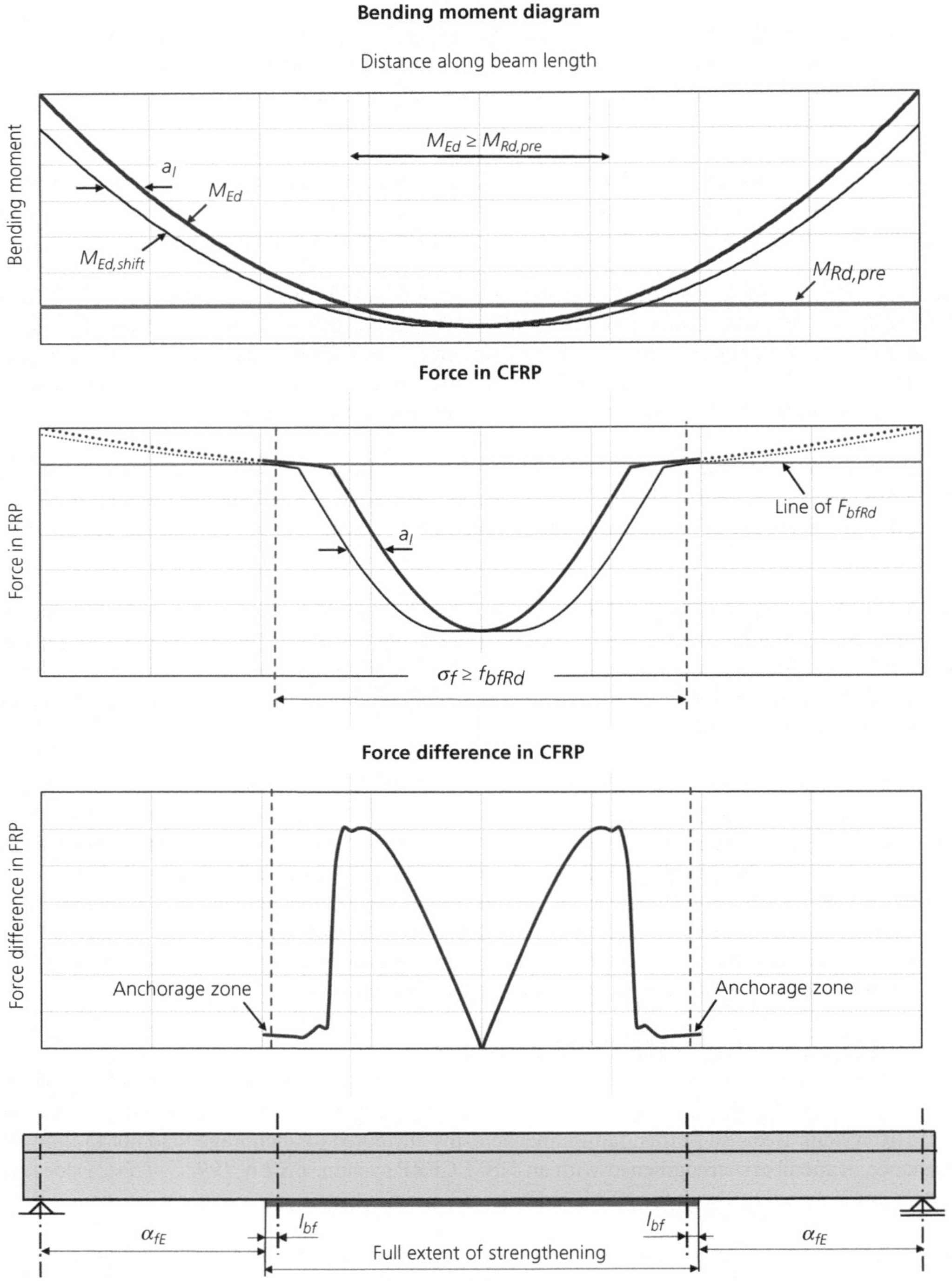

end cover separation and shear induced separation can occur. Relevant code checks need to be undertaken for these failure mechanisms to ensure they do not occur before the ULS resistance is achieved.

Moment redistribution

The CFRP strengthening system is not ductile. It cannot be relied upon to redistribute load as is common practice in structures reinforced with carbon steel or those which are pre- or post-tensioned. Therefore, redistribution or plastic analysis has been omitted as an option for structural analysis unless rigorous analysis is undertaken.

Examples of what could be considered in design are as follows.

- This provision only applies to regions of the structure which are strengthened so redistribution should not be discounted in the remainder of the structure.
- Analysis could be undertaken to prove rotation and ductility capacity for the required redistribution (redistribution may be undertaken as outlined in Tajaddini *et al.*, 2016).
- Calibrated non-linear finite element analysis could also be undertaken. Such an analysis would model the non-ductile response of the CFRP system alongside the ductile system comprising steel reinforcement, to redistribute action effects.

Where redistribution is undertaken, consideration of additional demand on the concrete-CFRP interface needs to be taken into account when assessing action effects for the bond. TR55 (Concrete Society, 2013) recommends that where moment redistribution capacity is proven, no more than 15% of the moment should be redistributed within the strengthened region. This is discussed in detail in Tajaddini *et al.* (2016).

Additional considerations for the accidental limit state

As stated in *BS EN 1992-1-1:2023*, 7(2), consideration needs to be given to accidental situations in design where the CFRP could become ineffective. This can include in a situation of loss of strength from fire damage in a building, or a where the CFRP system is substantially damaged such that it loses its strength, e.g. impact from a vehicle on a highway structure or vandalism.

The accidental limit state from BS EN 1990:2023 (BSI, 2023a) may be used in these situations to arrive at action effects in the structure, that can be used to determine utilisation with the partial material factors provided in *BS EN 1992-1-1:2023*, Table 4.3 and J.1. The CFRP strengthening may not be effective in these situations. *BS EN 1992-1-1:2023* requires that CFRP is not considered to be effective in the following situations.

- vandalism or accidental mechanical damage
- fire.

If it is identified through the design process that such cases may govern the design, additional strengthening may be required to supplement CFRP strengthening to avoid these situations arising.

The code also requires that the designer check that member collapse will not result in progressive collapse of a structure.

Example Flexural Design 1

Calculate the strengthened ULS capacity of the slab illustrated in Figure 5.5, assuming the material properties and reinforcement details, shown in Table 5.1, and that the slab is cast from the top relative to the tension reinforcement.

The slab has a span of 8 m, and a ULS uniformly distributed load (UDL) of 27.5 kN/m^2 and a UDL at strengthening of 6.25 kN/m^2.

CFRP properties.

The following design properties apply for the CFRP system.

$$f_{fud} = \eta_f \frac{f_{fuk}}{\gamma_f}$$

From *BS EN 1992-1-1:2023*, J.5.3(2) $\eta_f = 0.7$

From *BS EN 1992-1-1:2023*, Table J.1. $\gamma_f = 1.3$

Figure 5.5 Cross-sectional information for Example Flexural Design 1 – all dimensions are in mm unless stated otherwise

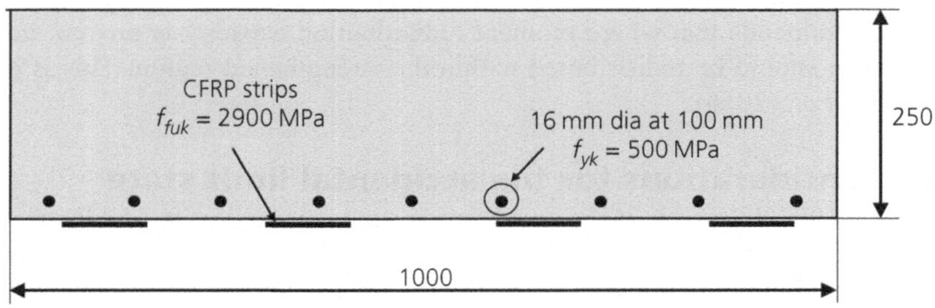

Table 5.1 Material properties for Example Flexural Design 1

Reinforcing Steel Properties		CFRP Properties		Concrete Properties	
d_{st}	192 mm	E_f	170 000 MPa	Grade	C35/45
ϕ	16 mm	f_{fuk}	2900 MPa	f_{cm}	43 MPa
Spacing	100 mm			f_{ck}	35 MPa
A_{st}	2010 mm^2/m			f_{cd}	19.8 MPa
f_{yk}	500 MPa			f_{ctm}	3.2 MPa
f_{yd}	435 MPa			$f_{ctk,0.05}$	2.2 MPa
E_s	200 000 MPa			f_{ctd}	1.0 MPa
ε_{uk}	5% (Ductility class B)			E_{cm}	33 282 MPa

Therefore

$$f_{fud} = 0.7 \times \frac{2900}{1.3} = 1562\,\text{MPa}$$

$$\varepsilon_{fud} = \frac{f_{fud}}{E_f} = \frac{1562}{170\,000} = 0.919\%$$

Flexural resistance

The flexural resistance may be found by using Equation 5.3 given below, with the strain limits also stated below.

$$M = \sigma_{st} A_{st} \left(d_{st} - k_2 x\right) + E_f A_f \varepsilon_f \left(h - k_2 x\right)$$

Where

$$k_1 = \begin{cases} 1000\varepsilon_c \left[0.5 - \dfrac{250\varepsilon_c}{3}\right], \varepsilon_c < 0.002 \\ 1 - \dfrac{1}{1500\varepsilon_c}, 0.002 \leq \varepsilon_c \leq 0.0035 \end{cases}$$

$$k_2 = \begin{cases} \dfrac{2 - 250\varepsilon_c}{6 - 1000\varepsilon_c}, \varepsilon_c < 0.002 \\ \dfrac{1000\varepsilon_c\left(3\,000\varepsilon_c - 4\right) + 2}{2\,000\varepsilon_c\left(3\,000\varepsilon_c - 2\right)}, 0.002 \leq \varepsilon_c \leq 0.0035 \end{cases}$$

The strain limits are determined as follows.

- The ultimate compressive strain of concrete as defined in *BS EN 1992-1-1:2023*, 8.1.2 $\left(\varepsilon_{cu} = 0.0035\right)$.
- The ultimate strain of reinforcing steel, pre-stressing or post-tensioning $\left(\varepsilon_{ud} = \dfrac{0.05}{1.15} = 0.043\right)$ per *BS EN 1992-1-1:2023*, Table C.4 assuming Class B reinforcement.
- The strain at the design tensile strength of the CFRP, (ε_{fud} = 0.00919), determined using Equation 2.2, when strengthened.

Flexural resistance pre-strengthening

Applying the relevant limits to the section pre-strengthening, the results in Table 5.2 are obtained.

The moment at ULS is calculated as

$$M_{Ed} = 27.5 \times \frac{8^2}{8} = 220\ \text{kNm/m}$$

The moment resistance is limited by concrete crushing, therefore

$$\text{M}_{Rd,pre} = 148\ \text{kNm/m} < \text{M}_{Ed} = 220\ \text{kNm/m}$$

Therefore, strengthening is needed.

Table 5.2 Pre-strengthening stress-strain analysis Example Flexural Design 1

Limit	ε_c	$\varepsilon_{st.1}$	x [mm]	M [kNm]
Concrete crushing ε_{cu} **= 0.0035**	0.0035	0.0089	54.4	148
Reinforcement rupture ε_{ud} **= 4.3%**	Not checked as it is apparent that concrete crushing occurs well ahead of this limit and will govern flexural resistance.			

Flexural resistance post-strengthening

Try $A_f = 320\ \text{mm}^2/\text{m}$ with the properties stated in the example.

Check if the cracking moment is greater than the moment at strengthening.

Calculate the cracking moment for the concrete tensile strength f_{ctm} (refer to *BS EN 1992-1-1:2023*, 3.2.1) using the expression

$$M_{cr} = f_{ctm}\frac{I}{h_{max}} = 3.2\times\frac{1.3021\times 10^9}{\frac{250}{2}}\times 10^{-6} = 33.3\ \text{kNm}/\text{m}$$

The moment at strengthening is calculated as

$$M_{Ed,str} = 6.25\times\frac{8^2}{8} = 50\ \text{kNm/m}$$

As the applied bending moment at strengthening is greater than the cracking moment, the soffit strain at the time of strengthening needs to be determined and taken into account in the CFRP design. It is calculated using cracked section properties. A detailed explanation of these terms is provided in Chapter 7.

$$x_{pre} = \frac{-A_{st}E_s + \sqrt{(A_{st}E_s)^2 + 2bA_{st}E_sE_{cm}d_{st}}}{bE_c}$$

$$= \frac{-2010\times 200\,000 + \sqrt{(2010\times 200\,000)^2 + 2\times 1000\times 2010\times 200\,000\times 33\,282\times 192}}{1000\times 33\,282}$$

$$x_{pre} = 57.1\ \text{mm}$$

$$I_{cr,pre} = \frac{bx_{pre}^{\ 3}}{3}\frac{E_{cm}}{E_s} + A_{st}\left(d_{st} - x_{pre}\right)^2 = \frac{1000\times 57.1^3}{3}\frac{33\,282}{200\,000} + 2010\times(192 - 57.1)^2$$

$$I_{cr,pre} = 46.9\times 10^6\ \text{mm}^4/\text{m}$$

$$\varepsilon_0 = \left[\frac{M_{Ed,str}\left(h - x_{pre}\right)}{I_{cr,pre}}\frac{E_{cm}}{E_s}\right]/E_{cm} = \left[\frac{50\times 10^6\times(250 - 57.1)}{46.9\times 10^6}\times\frac{33\,282}{200\,000}\right]/33\,282 = 0.001$$

Applying the three limits to capacity, Equation 5.3 yields the sectional moments given in Table 5.3.

The bending moment resistance is limited by concrete crushing, giving

$$M_{Rd} = 220 \text{ kNm / m} = M_{Ed}$$

Therefore, bending moment resistance with the proposed $A_f = 320 \text{ mm}^2/\text{m}$ is ok.

A typical supplier product (strips) that would fulfil the requirements is 6 No 50mm-wide CFRP strips with a thickness of 1.2 mm per metre width, giving a total of 360 mm²/m width of the slab. This results in a higher bending moment resistance, which is ok.

The resulting maximum centre-to-centre spacing will be 140 mm across the width, and the clear spacing will be approximately 90 mm. Spacing limits from *BS EN 1992-1-1:2023*, J.12 will not be an issue.

Additional checks will be undertaken to ensure that this is sufficient for bond and anchorage.

Table 5.3 Post-strengthening stress-strain analysis Example Flexural Design 1

Limit	ε_c	$\varepsilon_{st.1}$	ε_f	x [mm]	M [kNm]
Concrete crushing ε_{cu} **= 0.0035**	0.0035	0.0052	0.0068	77	220
CFRP rupture ε_f **= 0.00919**	0.0037	0.0070	0.00919	85	244
Reinforcement rupture ε_{ud} **= 0.043**	Not checked as it is apparent that concrete crushing and CFRP rupture occur well ahead of this limit and will govern flexural resistance.				

Anchorage

To calculate the anchorage resistance, f_{bfRd}, and the required anchorage length to generate this resistance, $l_{bf,max,k}$, the parameters $s_{cr,min}$ and $f_{ctm,surf}$ are required.

Calculate the cracking moment in order to determine the minimum crack spacing, $s_{cr,min}$.

Calculate the mean bond stress for ribbed bars assumed to be cast in good conditions ($\kappa_{vb1} = 1.0$), as determined by *BS EN 1992-1-1:2023*, Formula J.46 using

$$f_{fbsm} = 0.43 f_{cm}^{\frac{2}{3}} = 0.43 \times 43^{\frac{2}{3}} = 5.28 \text{ MPa}$$

Calculate the bond force per unit length of bars as per *BS EN 1992-1-1:2023*, Formula J.45 using

$$F_{fbsm} = \sum_{i=1}^{10} n_{s,i} \varnothing_i \pi f_{fbsm} = 10 \times 16 \times \pi \times 5.28 = 2653 \text{ N/mm}$$

Which gives the minimum crack spacing (*BS EN 1992-1-1:2023*, Formula J.44)

$$s_{cr,min} = 1.5\frac{M_{cr}}{0.85hF_{fbsm}} = 1.5\times\frac{33.3\times10^{6}}{0.85\times250\times2653} = 89\text{ mm}$$

Calculate $f_{ctm,surf}$ using *BS EN 1992-1-1:2023*, Formula J.34.

For the soffit concrete of a beam cast from the top

$$f_{ctm,surf} = f_{ctm}\left[0.6+0.3\left(\frac{f_{ck}}{60}-0.2\right)\right] = 2.3\text{ MPa}$$

Calculate anchorage resistance using the refined method to *BS EN 1992-1-1:2023*, Formula J.38.

Assume $k_{sys,b1} = k_{sys,b2} = k_{sys,b3} = 1$

$$s_{f0k} = 0.2k_{sys,b2} = 0.2$$

$$\tau_{f1k} = 0.37k_{sys,b1}\left(f_{cm}f_{ctm,surf}\right)^{0.5} = 0.37\times\left(43\times2.3\right)^{0.5} = 3.7\text{ MPa}$$

$$l_{bf,max,k} = \frac{2}{k_{sys,b3}}\sqrt{\frac{E_f t_f s_{f0k}}{\tau_{f1k}}} = 2\times\sqrt{\frac{170\,000\times1.2\times0.2}{3.7}} = 211\text{ mm}$$

$$f_{bfk,max} = \sqrt{\frac{E_f\tau_{f1k}s_{f0k}}{t_f}} = \sqrt{\frac{170\,000\times3.7\times0.2}{1.2}} = 322.8\text{ MPa}$$

$s_{cr,min} = 89\text{ mm} < l_{bf,max,k}$, therefore $l_{bf} = 89$ mm

$$f_{bfRd} = \frac{\sqrt{\eta_{cc}k_{tc}k_{tt}}}{\gamma_{BA}}f_{bfk,max}\frac{l_{bf}}{l_{bf,max,k}}\left(2-\frac{l_{bf}}{l_{bf,max,k}}\right)$$

$$f_{bfRd} = \frac{\sqrt{1.0\times0.85\times0.7}}{1.5}\times322.8\times\frac{89}{211}\times\left(2-\frac{89}{211}\right) = 110.2\text{ MPa}$$

$$F_{bfRd} = f_{bfRd}A_f = 110.2\times360\times10^{-3} = 39.7\text{ kN}$$

To determine the termination point of the anchorage, the force in the CFRP needs to be calculated. This is presented in Table 5.4.

Table 5.4 also includes additional items that will be used for intermediate crack debonding (including pre-strengthening and ULS moments at spacing of $s_{cr,min}$ along the length of the beam, starting at midspan). Equation 5.10 is used to determine CFRP forces with the CFRP pre-strain deducted where relevant. Equation 5.8 and Equation 5.11 are used to check if the reinforcement is at yield or not. The changes in stress along the length of the CFRP are determined as follows

$$\Delta f_{fEd} = \frac{\Delta F_{fEd}}{A_f}$$

Table 5.4 Tabulated data – Example Flexural Design 1 (continued on next page)

Distance [mm]	$M_{Ed,str}$ [kNm]	M_{Ed} [kNm]	$\sigma_{s,pre}$ [MPa]	σ_s [MPa]	F_s [kN]	F_{fEd} [kN]	f_{fEd} [MPa]	ΔF_{fEd} [kN]	Δf_{fEd} [MPa]
0	0	0	0	0	0	0	0		
84.00	2.08	9.14	5.98	24.34	48.93	7.32	20.33	7.32	20.33
173.00	4.23	18.62	12.17	49.57	99.63	14.90	41.39	7.58	21.07
262.00	6.34	27.88	18.22	74.22	149.17	22.31	61.97	7.41	20.58
351.00	8.39	36.92	24.13	98.28	197.55	29.54	82.07	7.23	20.10
440.00	10.40	45.74	29.90	121.77	244.76	36.60	101.68	7.06	19.61
529.00	12.35	54.34	35.52	144.68	290.80	43.49	120.81	6.89	19.13
618.00	14.26	62.73	41.01	167.00	335.68	50.20	139.45	6.71	18.64
707.00	16.11	70.90	46.35	188.75	379.39	56.74	157.61	6.54	18.16
796.00	17.92	78.85	51.54	209.92	421.94	63.10	175.29	6.36	17.68
885.00	19.68	86.58	56.60	230.51	463.32	69.29	192.48	6.19	17.19
974.00	21.39	94.10	61.51	250.52	503.54	75.31	209.18	6.01	16.71
1063.00	23.04	101.39	66.28	269.94	542.59	81.15	225.41	5.84	16.22
1152.00	24.65	108.47	70.91	288.79	580.47	86.81	241.15	5.67	15.74
1241.00	26.21	115.33	75.39	307.06	617.19	92.30	256.40	5.49	15.25
1330.00	27.72	121.98	79.74	324.75	652.74	97.62	271.17	5.32	14.77
1419.00	29.18	128.40	83.94	341.85	687.13	102.76	285.45	5.14	14.29
1508.00	30.59	134.61	88.00	358.38	720.35	107.73	299.26	4.97	13.80
1597.00	31.95	140.60	91.91	374.33	752.40	112.53	312.57	4.79	13.32
1686.00	33.27	146.37	95.68	389.70	783.29	117.15	325.41	4.62	12.83
1775.00	34.53	151.93	99.32	404.49	813.02	121.59	337.75	4.45	12.35
1864.00	35.74	157.27	102.80	418.70	841.58	125.86	349.62	4.27	11.86
1953.00	36.91	162.38	106.15	432.32	868.97	129.96	361.00	4.10	11.38
2042.00	38.02	167.29	109.35	435.00	874.35	172.95	480.42	42.99	119.42
2131.00	39.08	171.97	112.42	435.00	874.35	202.27	561.86	29.32	81.44
2220.00	40.10	176.43	115.34	435.00	874.35	230.23	639.52	27.96	77.66
2309.00	41.06	180.68	118.11	435.00	874.35	256.82	713.38	26.59	73.87
2398.00	41.98	184.71	120.75	435.00	874.35	282.05	783.46	25.23	70.08
2487.00	42.85	188.52	123.24	435.00	874.35	305.91	849.76	23.87	66.29
2576.00	43.66	192.12	125.59	435.00	874.35	328.41	912.26	22.50	62.50
2665.00	44.43	195.49	127.79	435.00	874.35	349.55	970.98	21.14	58.72
2754.00	45.15	198.65	129.86	435.00	874.35	369.32	1025.90	19.77	54.93
2843.00	45.82	201.59	131.78	435.00	874.35	387.74	1077.04	18.41	51.14
2932.00	46.44	204.32	133.56	435.00	874.35	404.78	1124.39	17.05	47.35
3021.00	47.00	206.82	135.20	435.00	874.35	420.46	1167.96	15.68	43.56

Table 5.4 Continued

Distance [mm]	$M_{Ed,str}$ [kNm]	M_{Ed} [kNm]	$\sigma_{s,pre}$ [MPa]	σ_s [MPa]	F_s [kN]	F_{fEd} [kN]	f_{fEd} [MPa]	ΔF_{fEd} [kN]	Δf_{fEd} [MPa]
3110.00	47.52	209.11	136.69	435.00	874.35	434.78	1207.73	14.32	39.78
3199.00	47.99	211.18	138.05	435.00	874.35	447.74	1243.72	12.96	35.99
3288.00	48.42	213.03	139.26	435.00	874.35	459.33	1275.92	11.59	32.20
3377.00	48.79	214.66	140.33	435.00	874.35	469.56	1304.33	10.23	28.41
3466.00	49.11	216.08	141.25	435.00	874.35	478.42	1328.95	8.86	24.62
3555.00	49.38	217.28	142.03	435.00	874.35	485.92	1349.79	7.50	20.83
3644.00	49.60	218.26	142.68	435.00	874.35	492.06	1366.83	6.14	17.05
3733.00	49.78	219.02	143.17	435.00	874.35	496.83	1380.09	4.77	13.26
3822.00	49.90	219.56	143.53	435.00	874.35	500.24	1389.56	3.41	9.47
3911.00	49.98	219.89	143.74	435.00	874.35	502.29	1395.24	2.05	5.68
4000.00	50.00	220.00	143.81	435.00	874.35	502.97	1397.14	0.68	1.89

Curtailment needs to take account of a_l, calculated in accordance with *BS EN 1992-1-1:2023*, 12.3.2. For steel-reinforced members without shear reinforcement

$$a_l = d_{st}$$

Calculate the equivalent effective depth for the section with CFRP strengthening, using Equation 5.6.

$$d_{equiv} = \frac{d_f E_f A_f + \sum_{i=1}^{n} d_{st.i} E_s A_{st.i}}{E_f A_f + \sum_{i=1}^{n} E_s A_{st.i}} = \frac{250 \times 170\,000 \times 360 + 192 \times 200\,000 \times 2010}{170\,000 \times 360 + 200\,000 \times 2010}$$

$$d_{equiv} = 200 \text{ mm} = a_l$$

From this, it is evident that the CFRP force is less than F_{bfRd} at a distance of approximately 490 mm from the support. Adjusting for the shift term (increasing the length of CFRP by $a_l = 200$ mm at both ends of the beam), and adding the anchorage length ($l_{bf} = 89\,\text{mm}$) to both ends of the strengthening, the overall CFRP strengthening length will be

$$(8 - 0.49 \times 2) + 2 \times (0.089 + 0.2) = 7.6\text{m}$$

That is, the CFRP will be terminated at a distance of $\alpha_{fE} \approx 200$ mm from centre of support.

Note that in this case, using the value of d_{st} without considering the CFRP would not have made an appreciable difference on the location of the termination point of the CFRP. In this instance, the start of curtailment (once the shift rule is considered) is in the uncracked region of the member, therefore no additional anchorage length has been provided to ensure that the CFRP will extend beyond a full crack spacing.

Intermediate crack debonding

The slab has a sagging curvature and, therefore, all components of bond resistance in *BS EN 1992-1-1:2023*, Formula J.47 may be used.

$$\Delta f_{fk,B} = 0.84k_{sys,b1}\sqrt{f_{cm}f_{ctm,surf}}\frac{\sqrt{s_{cr,min}}}{t_f} = 0.84\times1\times\sqrt{43\times2.3}\times\frac{\sqrt{89}}{1.2} = 65.4\,\text{MPa}$$

$$\Delta f_{fk,F} = f_{cm}^{-0.9}\frac{s_{cr,min}^{\frac{4}{3}}}{t_f} = 43^{-0.9}\times\frac{89^{\frac{4}{3}}}{1.2} = 11.2\text{ MPa}$$

$$\Delta f_{fk,C} = \frac{\kappa_h}{h_f}\frac{s_{cr,min}^{\frac{1}{3}}}{t_f} = \frac{2000}{250}\times\frac{89^{\frac{1}{3}}}{1.2} = 29.7\text{ MPa}$$

$$\Delta f_{fRd} = \frac{1}{1.5}\times\left[\left(1.0\times0.85\times0.7\right)^{0.5}\times65.7+11.2+29.7\right] = 34+7.4+19.8 = 60.9\text{ MPa}$$

In determining the forces in the CFRP, and for use of Equation 5.10, the inner lever arm of forces is determined from Equation 5.7 as

$$z_m = 0.8d_{equiv}$$

$$z_m = 160\text{ mm}$$

As is evident from the highlighted cells in Table 5.4, the intermediate crack debonding resistance is lower than Δf_{fEd} for a distance of a little over 0.5 m both sides of the centre of the span of the beam. Further analysis, not presented here, indicates that this is in the area where the reinforcement yields.

Therefore, additional CFRP is needed to reduce stresses in the intermediate crack debonding check so they are less than the resistance between cracks.

Additional analysis also shows that using $A_f = 540\text{ mm}^2/\text{m}$, including 3 No 150mm-wide strips with $t_f = 1.2$ mm will be required.

In this instance, the anchorage resistance increases to $F_{bfRd} = 59.5$ kN, meaning that the length of the CFRP can be reduced, noting that without considering the shift term, the termination point would be approximately 534 mm from the support. The anchorage therefore begins within the cracked region, and an anchorage length of $2s_{cr,min}$ is provided, which is 178 mm. Provision of an anchor equal to $l_{bf,max,k}$ is not considered in this instance as the distance of $2s_{cr,min}$ will ensure that the distance of $l_{bf} = 89$ mm is provided outside of the cracked region. In this case, the length of the CFRP will be

$$(8-0.534\times2)+2\times(0.178+0.2) = 7.688\,\text{m}$$

With

$$\alpha_{fE} \approx 0.156\,\text{m}$$

End cover separation check (*BS EN 1992-1-1:2023*, J.11.1.2.4)

Calculate the maximum shear force at the end of CFRP reinforcement at distance $\alpha_{fE} = 0.129$ m from centre of support as follows

$$V_{Ed} = 27.5 \times \frac{8}{2} - 27.5 \times 0.129 = 106 \text{ kN/m}$$

$$\rho_l = \frac{A_{sl}}{b_w d_{st}} = \frac{2010}{1000 \times 192} = 0.0105 \text{ (\textit{BS EN 1992-1-1:2023}, Formula 8.28)}$$

$$V_{Rd,cfE} = \left(0.11 + 2.2 \frac{(100\rho_l)^{0.15}}{\alpha_{fE}^{0.36}} \right) (100\rho_l f_{ck})^{\frac{1}{3}} b_w d_{st}$$

$$V_{Rd,cfE} = \left(0.11 + 2.2 \times \frac{(100 \times 0.0105)^{0.15}}{129^{0.36}} \right) \times (100 \times 0.0105 \times 35)^{\frac{1}{3}} \times 1000 \times 192$$

$$V_{Rd,cfE} = 316 \text{ kN/m}$$

$$V_{Ed} = 106 \text{ kN/m} < V_{Rd,cfE} = 316 \text{ kN/m}$$

End cover separation checks are satisfied and no additional strengthening is needed.

Shear induced separation check (*BS EN 1992-1-1:2023*, J.11.1.2.5)

To check shear crack induced separation, the maximum shear throughout the full length of the CFRP is used, which in this case is at the end of the anchorage, closest to the support.

$$V_{Ed} = 106 \text{kN / m}$$

Calculate the design shear stress at the end of CFRP reinforcement in accordance with *BS EN 1992-1-1:2023*, Formula 8.18.

$$\tau_{Ed} = \frac{V_{Ed}}{b_w z} = \frac{V_{Ed}}{b_w 0.9 d_{st}} = \frac{106 \times 10^3}{1000 \times 0.9 \times 192} = 0.61 \text{ MPa}$$

Check in accordance with *BS EN 1992-1-1:2023*, Formula J.54.

$$\tau_{Ed} = 0.61 \text{ MPa} < 0.33 f_{ck}^{\frac{2}{3}} = 0.33 \times 35^{\frac{2}{3}} = 3.5 \text{ MPa} - \text{check is ok.}$$

Conservatively, take $cot\theta = 1$ and calculate the shear resistance as the upper limit of *BS EN 1992-1-1:2023*, Formula 8.62.

$$\tau_{Rd} = \nu f_{cd} \frac{cot\theta}{1 + cot^2\theta} = 0.5 \times 19.8 \times \frac{1.0}{1 + 1.0^2} = 4.95 \text{ MPa}$$

Conservatively,

$$\sigma_{swd} = f_{ywd} = 435 \text{ MPa}$$

For ribbed steel bars (*BS EN 1992-1-1:2023*, Formula J.53) the following condition needs to be satisfied.

$$\frac{\tau_{Ed}\sigma_{swd}}{\tau_{Rd}} = \frac{0.61 \times 435}{4.95} = 53.6\,\text{MPa} < 75\,\text{MPa}$$

Shear-induced checks are satisfied and no additional strengthening is needed.

Example Flexural Design 2

As stated in the chapter text, the CFRP force differences calculated using iterative methods for intermediate crack debonding checks are often lower than the simplified method used in Example Flexural Design 1. This example (Example Flexural Design 2) implements an iterative approach for all aspects of the check to illustrate the benefit of using the strain compatibility-based iterative method.

Calculate the strengthening requirements for the slab in Example Flexural Design 1 using the detailed iterative procedure outlined in Annex A for internal cross-sectional forces. In this case, 80mm-wide CFRP strips are used. Example Flexural Design 1 has indicated that 4 of these strips will be sufficient for flexural resistance of the section, as shown in Figure 5.6.

A computer script was written to undertake the detailed calculation of section forces and resistances based on parabola-rectangle stress distribution in the compression zone given in *BS EN 1992-1-1:2023*, Figure 8.2c, using the iterative approach described in Annex A.

Using this code, the resistance of the existing section pre-strengthening was found to be

$$M_{Rd,pre} = 148 \text{ kNm/m} < M_{Ed} = 220 \text{ kNm/m}$$

$\varepsilon_{cu} = 0.0035 = 0.35\%$ (*BS EN 1992-1-1:2023*, Formula 8.4)

Neutral axis: $x = 54.4$ mm

Figure 5.6 Cross-sectional information for Example Flexural Design 2 – all dimensions are in mm unless stated otherwise

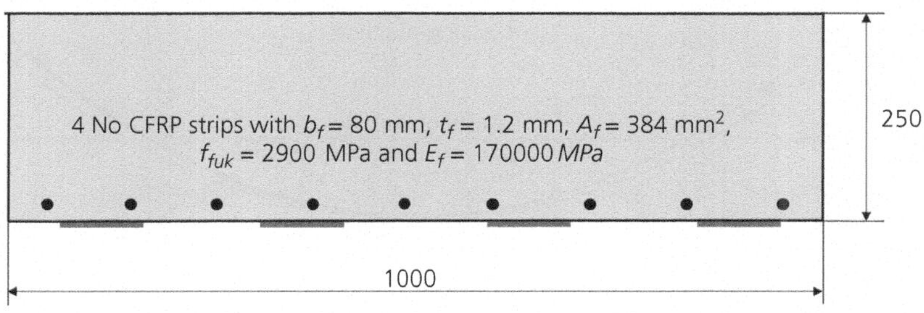

Therefore, the capacity of the existing section is not adequate for the applied moment at ULS and strengthening is required, as evident from Figure 5.7.

Calculate strains in steel and concrete at strengthening

The strengthening is applied under permanent loads – that is

$$M_{Ed,str} = 50 \text{ kNm/m}$$

The stress-strain analysis is performed using the parabola-rectangle stress distribution in the compression zone given in *BS EN 1992-1-1:2023*, Figure 8.2c, applying the iterative approach described in Annex A.

$$x_{pre} = 72.4 \text{ mm}$$

Top surface strain

$$\varepsilon_{c,pre} = 0.00045$$

Soffit surface strain

$$\varepsilon_0 = 0.00110$$

Calculate the ULS capacity of the strengthened section

Assuming concrete fails first based on parabola-rectangle stress distribution in the compression zone given in *BS EN 1992-1-1:2023*, Figure 2c (using the iterative approach described in Annex A)

$$x = 80.1 \text{ mm}$$

$$M_{Rd} = 228.0 \text{ kNm / m}$$

Check strain in CFRP

$$\varepsilon_f = 0.00632 = 0.632\% < \varepsilon_{fud} = 0.919\%$$

Figure 5.7 Applied and resistance bending moments including shift diagram for Example Flexural Design 2

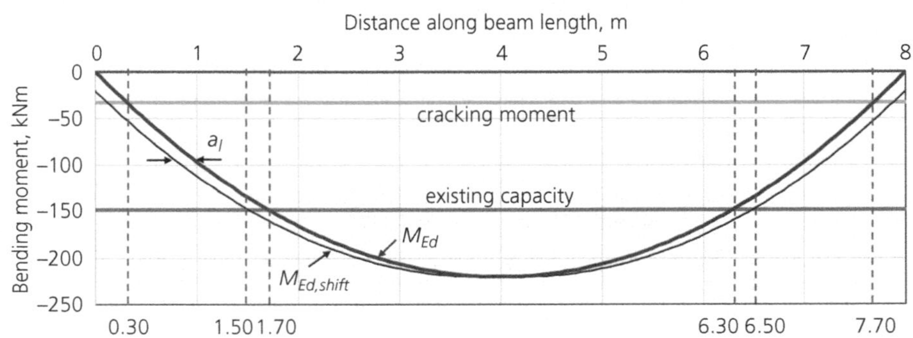

Therefore, the ULS moment capacity of the strengthened section is limited by crushing of concrete.

Check maximum steel strain

$$\varepsilon_{st} = 0.00489 = 0.489\% < \frac{5\%}{1.15} = 4.3\%$$

End anchorage (*BS EN 1992-1-1:2023*, J.11.1.2.2)

The values of $l_{bf,max,k}$ and f_{bfRd} are identical for this example as in Example Flexural Design 1, and are

$$l_{bf,max,k} = 211 \text{ mm}$$

$$f_{bfk,max} = 322.8 \text{ MPa}$$

$$f_{bfRd} = 110.2 \text{ MPa}$$

$$F_{bfRd} = f_{bfRd} A_f = 110.2 \times 384 = 42.3 \text{ kN}$$

For members without shear reinforcement (*BS EN1992-1-1:2023*, Formula 12.6) and as suggested in Flexural Example 1, the shift distance may be taken for simplicity as $a_l = d_{st} = 192$ mm

As indicated by the results presented in Figure 5.8, the overall CFRP strengthening length is $(7.718 - 0.282) + 2 \times 0.089 = 7.614$ m (at $\alpha_{fE} = 0.193$ m distance from centre of support).

Figure 5.8 Force in CFRP and end anchorage resistance for Example Flexural Design 2

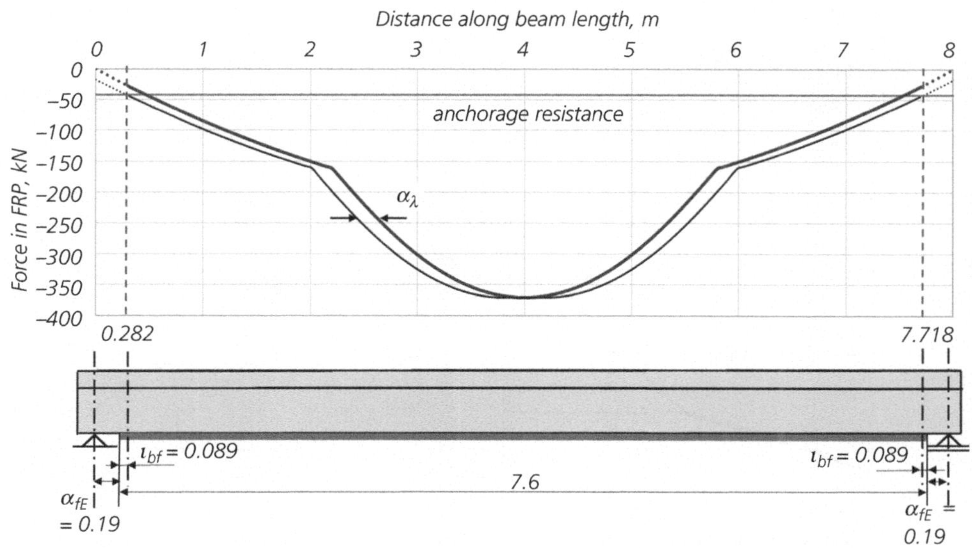

As can be observed from Figure 5.7 and Figure 5.8, the anchorage length in this occasion falls outside the cracked zone. Therefore, it is safe to provide $l_{bf} = 0.089$ m.

Re-calculating the anchorage resistance for $l_{bf,max,k} = 211$ mm gives $F_{bfRd} = 63.8$ kN > $F_{Ed} = 44.5$ kN at the location of M_{cr}. Providing overall length of (8−2 × 0.3) + 2 × 0.211 = 7.822 m > 7.614 m is not resulting in more efficient design. However, fine-tuning the required anchorage length outside the cracked zone for a force in CFRP of $F_{Ed} = 44.5$ kN results in $l_{bf} = 95$ mm, reducing the overall required length to (8−2 × 0.3) + 2 × 0.095 = 7.59 m ~ 7.6 m, which indicates that both approaches result in similar CFRP required lengths.

Intermediate crack debonding (*BS EN 1992-1-1:2023*, J.11.1.2.3)

As in Example Flexural Design 1

$$\Delta f_{fRd} = 60.9 \text{ MPa}$$

Maximum $\Delta f_{fEd} = 47.2$ MPa < Δf_{fRd} as illustrated in Figure 5.9.

Tabulated results are also given in Table 5.5.

End cover separation check (*BS EN 1992-1-1:2023*, J.11.1.2.4)

Calculate the maximum shear force at the end of CFRP reinforcement calculated at distance $\alpha_{fE} = 0.193$ m from centre of support as follows

$$V_{Ed} = 104.7 \text{ kN/m}$$

$$\rho_l = \frac{A_{sl}}{b_w d_{st}} = \frac{2010}{1000 \times 192} = 0.0105 \text{ (\textit{BS EN 1992-1-1:2023}, Formula 8.28)}$$

$$V_{Rd,cfE} = \left(0.11 + 2.2\frac{(100\rho_l)^{0.15}}{\alpha_{fE}^{0.36}}\right)(100\rho_l f_{ck})^{\frac{1}{3}} b_w d_{st}$$

Figure 5.9 Force in CFRP and steel (a) and force difference in CFRP (b) for Example Flexural Design 2

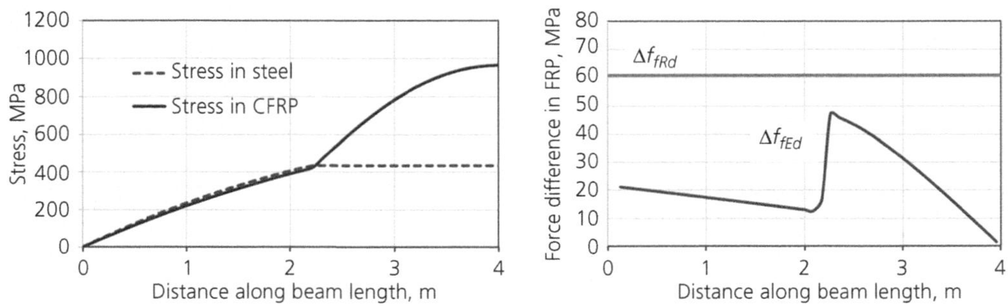

Table 5.5 Tabulated data – Example Flexural Design 2 (continued on next page)

Distance [m]	M_{Ed} [kNm]	V_{Ed} [kN]	x [mm]	a_l [mm]	F_{fEd} [kN]
0.00	0	110	76.3	192	
0.08	9	108	75.1	192	
0.17	19	105	75.5	192	15.9
0.26	28	103	75.8	192	23.9
0.35	37	100	76.2	192	31.7
0.44	46	98	76.6	192	39.4
0.53	54	95	77.0	192	47.0
0.62	63	93	77.4	192	54.3
0.71	71	91	77.8	192	61.6
0.80	79	88	78.2	192	68.7
0.89	87	86	78.6	192	75.6
0.97	94	83	79.0	192	82.4
1.06	101	81	79.4	192	89.1
1.15	108	78	79.8	192	95.6
1.24	115	76	80.2	192	102.0
1.33	122	73	80.6	192	108.2
1.42	128	71	81.0	192	114.2
1.51	135	69	81.5	192	120.2
1.60	141	66	81.9	192	125.9
1.69	146	64	82.3	192	131.5
1.78	152	61	82.8	192	137.0
1.86	157	59	83.2	192	142.3
1.95	162	56	83.7	192	147.5
2.04	167	54	84.1	192	152.5
2.13	172	51	84.6	192	157.3
2.22	176	49	84.7	192	163.8
2.31	181	47	83.3	192	181.9
2.40	185	44	82.2	192	199.5
2.49	189	42	81.4	192	216.7
2.58	192	39	80.7	192	233.2
2.67	195	37	80.2	192	249.0
2.75	199	34	79.9	192	264.1
2.84	202	32	79.6	192	278.3
2.93	204	29	79.5	192	291.6

Table 5.5 Continued

Distance [m]	M_{Ed} [kNm]	V_{Ed} [kN]	x [mm]	a_l [mm]	F_{fEd} [kN]
3.02	207	27	79.4	192	303.9
3.11	209	24	79.3	192	315.3
3.20	211	22	79.3	192	325.7
3.29	213	20	79.3	192	335.0
3.38	215	17	79.3	192	343.3
3.47	216	15	79.4	192	350.5
3.56	217	12	79.4	192	356.6
3.64	218	10	79.4	192	361.6
3.73	219	7	79.5	192	365.5
3.82	220	5	79.5	192	368.3
3.91	220	2	79.5	192	370.0
4.00	220	0	79.5	192	370.6

$$V_{Rd,cfE} = \left(0.11 + 2.2 \times \frac{(100 \times 0.0105)^{0.15}}{193^{0.36}}\right) \times (100 \times 0.0105 \times 35)^{\frac{1}{3}} \times 1000 \times 192$$

$$V_{Rd,cfE} = 282\,584\ N = 282.6\ \text{kN/m}$$

$$V_{Ed} = 104.7\ \text{kN/m} < V_{Rd,cfE} = 282.6\ \text{kN/m} - \text{check is ok.}$$

Shear induced separation check (*BS EN 1992-1-1:2023*, J.11.1.2.5)

$$V_{Ed} = 104.7\ \text{kN / m}$$

Calculate the design shear stress at the end of CFRP reinforcement in accordance with *BS EN 1992-1-1:2023*, Formula 8.18.

$$\tau_{Ed} = \frac{V_{Ed}}{b_w z} = \frac{V_{Ed}}{b_w 0.9 d_{st}} = \frac{104.7 \times 10^3}{1000 \times 0.9 \times 192} = 0.61\ \text{MPa}$$

Check in accordance with *BS EN 1992-1-1:2023*, Formula J.54.

$$\tau_{Ed} = 0.61\ \text{MPa} < 0.33 f_{ck}^{\frac{2}{3}} = 0.33 \times 35^{\frac{2}{3}} = 3.5\ \text{MPa} - \text{check is ok.}$$

Conservatively, take $cot\theta = 1$ and calculate the shear resistance as the upper limit of *BS EN 1992-1-1:2023*, Formula 8.62.

$$\tau_{Rd} = \nu f_{cd} \frac{cot\theta}{1+cot^2\theta} = 0.5 \times 19.8 \times \frac{1.0}{1+1.0^2} = 4.95 \text{ MPa}$$

Conservatively,

$$\sigma_{swd} = f_{ywd} = 435 \text{ MPa}$$

For ribbed steel bars (*BS EN 1992-1-1:2023*, Formula J.53)

$$\frac{\tau_{Ed}\sigma_{swd}}{\tau_{Rd}} = \frac{0.61 \times 435}{4.95} = 54 \text{ MPa} < 75 \text{ MPa} \text{ – check is ok.}$$

Design with strut and tie models and stress fields

Annex J of *BS EN 1992-1-1:2023* permits the use of CFRP strengthening in ties as part of a strut and tie system. The only single provision that is required to be met is that strain compatibility is achieved.

Example Strut and Tie: Pier strengthening

The bridge pier illustrated in Figure 5.10 has inadequate existing reinforcement required to carry the force in tie 1. CFRP NSM reinforcement is provided at each face to enhance the tensile capacity.

Loading

Additional force to be taken by CFRP strengthening: 165 kN

Per face: $\frac{165}{2} = 82.5$ kN

Figure 5.10 Example Strut and Tie strengthening of bridge pier CFRP NSM strengthening details – all dimensions are in mm unless stated otherwise

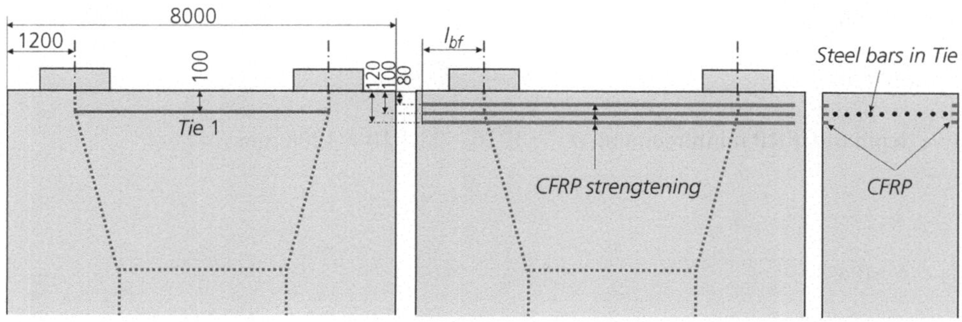

Materials

Concrete

Characteristic compressive strength

$$f_{ck} = 30 \text{ MPa}$$

$$D_{upper} = 20 \text{ mm}$$

CFRP strips properties

$$f_{fud} = 1562 \text{ MPa}$$

Thickness of strip: $t_f = 2.5$ mm

Width of strip: $b_f = 20$ m

Area per NSM Strip 50mm^2

For three strips

$$A_f = 3 \times 50 = 150 \text{ mm}^2$$

BS EN 1992-1-1:2023, Table J.2 requires the following.

- Edge distances, $a_r \geq 4\ b_f$, therefore $a_r \geq 80$ mm
- Slot width $t_f + 2 \leq b_{slot} \leq t_f + 4$, take $b_{slot} = 5$ mm
- Slot depth $b_f \leq t_{slot} \leq b_f + 2$ take $t_{slot} = 22$ mm
- Centre-to-centre spacing for a non-flexural member $s_f \geq max\left\{b_f, D_{upper}\right\}$

To ensure that the strengthening is centred on the same line as the reinforcement, and because the minimum edge distance is 80 mm, a value of $s_f = 20$ mm is selected for use.

$$a_{r,1} = 80 \text{ mm}$$

$$a_{r,2} = 100 \text{ mm}$$

$$a_{r,3} = 120 \text{ mm}$$

Effective depth of CFRP reinforcement: $d_f = 1220 - 22 + 10 = 1208$ mm

Adhesive properties

$$\tau_{bAd} = 6 \text{ MPa}$$

Check tensile capacity of CFRP reinforcement

$$F_{fRd} = 150 \times 1562 = 234.3\ \text{kN} > 82.5\ \text{kN per strip}$$

Check anchorage force resistance of CFRP reinforcement

For each NSM strip

$$F_{bfRd,i} = 0.95 b_f \tau_{bAd} a_r^{\frac{1}{4}} \left(26.2 + 0.065 tanh\left(\frac{a_r}{70}\right)\left(l_{bf} - 115\right)\right)$$

Take $l_{bf} = 1015\ \text{mm} > 115\ \text{mm}$

$$F_{bfRd,1} = 0.95 \times 20 \times 6 \times 80^{\frac{1}{4}} \times \left(26.2 + 0.065 tanh\left(\frac{80}{70}\right)(1015 - 115)\right) = 25195\ \text{N}$$

$$F_{bfRd,2} = 0.95 \times 20 \times 6 \times 100^{\frac{1}{4}} \times \left(26.2 + 0.065 tanh\left(\frac{100}{70}\right)(1015 - 115)\right) = 28\,244\ \text{N}$$

$$F_{bfRd,3} = 0.95 \times 20 \times 6 \times 120^{\frac{1}{4}} \times \left(26.2 + 0.065 tanh\left(\frac{120}{70}\right)(1015 - 115)\right) = 30\,572\ \text{N}$$

$$F_{bfRd} = F_{bfRd,1} + F_{bfRd,2} + F_{bfRd,3} = 25195 + 28\,244 + 30\,572 = 84\,010\ \text{N} = 84\ \text{kN} > 82.5\ \text{kN}$$

Therefore, three CFRP NSM strips 2.5 mm by 20 mm per face would be adequate, when anchored 1015 mm beyond the centreline of bearing plinths.

Kaloyana Kostova and Craig Giaccio
ISBN 978-1-83549-935-1
https://doi.org/10.1108/978-1-83549-934-420251006

Chapter 6
Columns

Introduction

The beneficial effect of confinement that is produced by wrapping columns with FRP is a well-established strengthening method. Strengthening is undertaken by wrapping a column with CFRP in accordance with *BS EN 1992-1-1:2023* (BSI, 2023b) either as a fully wrapped system encasing the column from top to bottom or a partially wrapped system. Figure 6.1 shows a fully wrapped CFRP system applied to a column. Member compression resistances can be increased using this method, as can their ductility.

Confinement

BS EN 1992-1-1:2023 covers the following applications of column strengthening with CFRP.

- Strengthening with continuous in situ lay-up CF sheets.
- Strengthening with CFRP strips orientated at an angle β_f to the axis perpendicular to the direction of the compression force in a column, as illustrated in Figure 6.2.

As the concrete compressive stress increases on a member in compression, the strain also increases. This results in lateral strains – that is, a small amount of expansion of the column across its cross section. When a column is strengthened with a CFRP wrapping system with fibres orientated perpendicular to the longitudinal axis of the column, the CFRP system resists these expansion strains and produces a confining stress around the cross section which, in turn, produces a triaxial stress state on the concrete. This triaxial stress state increases the compressive strength of the concrete, thereby increasing the compression resistance of the column. Wrapping also results in an increase of the ultimate strain that can be exhibited by a concrete column. An illustration of this confinement mechanism is shown in Figure 6.3.

The magnitude of the confining stress that the CFRP can apply to a concrete column is directly related to the stiffness of the CFRP wrap. This is governed by the modulus of elasticity, design rupture strength, the thickness of the wrap and how the wrap is applied (e.g. discrete strips, helical strips or fully wrapped). The constitutive model used to describe the stress-strain response of the confined concrete material, incorporating these parameters, is outlined later in this chapter. Provisions in *BS EN 1992-1-1:2023* were not written for seismic situations. These are covered in BS EN 1998-1:2004 (BSI, 2004d).

The confining effect of CFRP is much more effective in circular columns than in rectangular columns. This is because the confining effect is evenly distributed around the perimeter of a circular column, whereas the confinement is concentrated at corners on square or rectangular columns. With respect to rectangular sections, the confining effect is most effective on square sections, with the confinement effect becoming less and less effective as the aspect ratio increases. The latter is illustrated in Figure 6.3.

Figure 6.1 Illustration of full height column wrapping with CFRP (Photo courtesy of Concrete Repairs Limited (UK)

Limitations on applications where confinement from CFRP can be considered in the code provisions are provided in *BS EN 1992-1-1:2023*, J.8.1.2, as outlined below. These are mostly introduced because they reflect limitations in the testing databases used for development of the code provisions, as well as the required level of reliability prescribed by the Eurocode system.

- The characteristic compressive strength of columns, f_{ck}, shall be less than 50 MPa.
- A minimum diameter of $D_{eq} \geq 150$ mm applies for circular columns. This limit also applies to rectangular columns, using the effective diameter calculated in Equation 6.1 (Formula J.11).
- The first order eccentricity satisfies the provision in Equation 6.2 to ensure that the CFRP does not produce a confining effect where tension is dominant in the section, which is discussed in detail later in this chapter.
- The slenderness is limited to 40 as stated in Equation 6.3.

Figure 6.2 Definition of geometric symbols for confinement provisions (Copied from *BS EN 1992-1-1:2023*, Figure J.2)

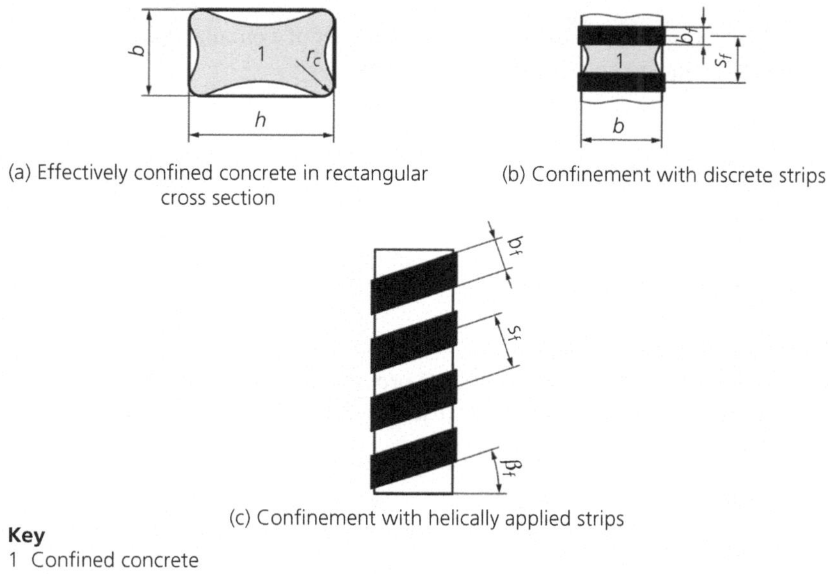

Figure 6.3 Confinement mechanism and illustration of strengthening with CFRP strips (Adapted from TR55 (Concrete Society, 2013))

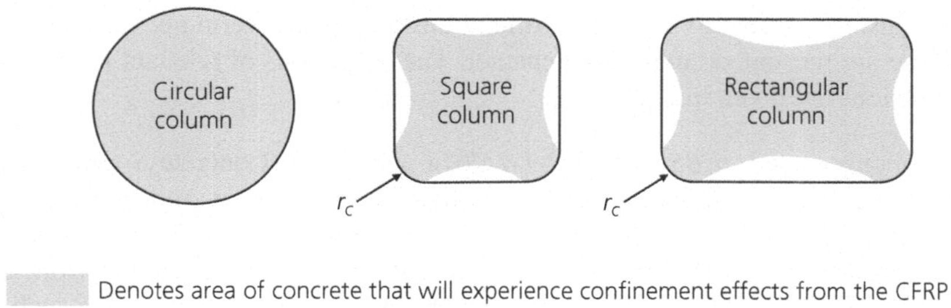

- A corner radius $r_c \geq 20$ mm is provided. This limit is provided because CFRP is not effective on columns with a corner radius less than this value.
- Provisions in this standard for rectangular columns should only be applied where the aspect ratio $\left(\frac{h}{b}\right) < 2$.

$$D_{eq} = \frac{2bh}{b+h} \tag{6.1}$$

$$\frac{e_0}{D_{eq}} \leq 0.2 \tag{6.2}$$

Where

e_o is the first order eccentricity.

D_{eq} is the equivalent column diameter, which is the diameter of a circular column, D, or is calculated using Equation 6.1 for square and rectangular columns.

$$\frac{l_0}{D_{eq}} \leq 40 \tag{6.3}$$

Where

l_o is the effective length of the column.

General formulation

The provisions in *BS EN 1992-1-1:2023*, J.8.1.2 provide a value of Δf_{cd} for use in Equation 6.4, which is compatible with *BS EN 1992-1-1:2023*, Formula 8.15 when considering confinement provided by stirrups.

$$f_{cd,c} = f_{cd} + \Delta f_{cd} \tag{6.4}$$

Where

$f_{cd,c}$ is the design strength of concrete confined by CFRP

Δf_{cd} is the increase in design strength that results from wrapping with a CFRP system.

Given that CFRP wrapping is much more effective for circular columns than for square or rectangular columns, the code provides two formulations for Δf_{cd}, covering circular and rectangular columns. These are presented later in this chapter. The definitions of relevant geometric symbols for these provisions are given in Figure 6.2.

While not explicitly stated in *BS EN 1992-1-1:2023* and TR55 (Concrete Society, 2013), CFRP confinement is considered without the beneficial effect of confinement from transverse steel reinforcement. This is because there is no clear model that describes the interaction between column confinement reinforcement and CFRP confinement. Additionally, experimental work to quantify the confining effect of CFRP has often utilised lower values of column links than such that might produce a real confining effect, meaning that there is no reliable experimental indication of how these two confinement mechanisms work together. For these reasons, superimposing the confining effect from CFRP onto that of confinement reinforcement is not applied in this guide. Note also that ACI 440.2R (ACI, 2017) does not include a superposition of these two contributions to confinement.

Circular columns

For circular columns, the additional confining resistance is calculated using code provisions as shown in Equation 6.5.

$$\Delta f_{cd} = \begin{cases} 0 & \text{for } \dfrac{t_f k_h f_{fud}}{D f_{cd}} < 0.07 \\ k_{cc} \dfrac{t_f}{D} k_h f_{fud} & \text{for } \dfrac{t_f k_h f_{fud}}{D f_{cd}} \geq 0.07 \end{cases} \tag{6.5}$$

Where

k_{cc} is the confinement factor for columns taken as $k_{cc} = 2.5$ for circular columns.

k_h is a factor applied for systems that are wrapped intermittently or helically. For clarity the factor has been applied in Equation 6.5 differently to *BS EN 1992-1-1:2023*, Formulae J.6 and J.7 and should be taken as shown in Equation 6.6.

$$k_h = \begin{cases} \left(1 - \frac{(s_f - b_f)}{2D}\right)^2 \left(\frac{1}{1+(tan\beta_f)^2}\right) & \text{for helical and/or intermittently wrapped systems} \\ 1 & \text{otherwise} \end{cases} \quad (6.6)$$

The provision for circular columns in the code was taken from fib Bulletin No. 90 (fib, 2019) and manipulated to achieve the desired consistency with other provisions in *BS EN 1992-1-1:2023*. The strength enhancement factor k_{cc} has been reduced in the code to a value of 2.5 from 3.3 in fib Bulletin No. 90 (fib, 2019) to reflect the fact that a characteristic value is needed for design, as described in the Background Document to FprEN 1992-1-1 (European Committee for Standardisation, 2023).

Square or rectangular columns

For square or rectangular columns, the additional confining resistance is calculated using code provisions as shown in Equation 6.7.

$$\Delta f_{cd} = \begin{cases} 0 & \text{for} \left(\frac{b}{h}\right)^2 k_e \frac{t_f k_r k_h f_{fud}}{D_{eq} f_{cd}} < 0.07 \\ k_{cc} \left(\frac{b}{h}\right)^2 k_e \left(\frac{t_f}{D_{eq}}\right) k_r k_h f_{fud} & \text{for} \left(\frac{b}{h}\right)^2 k_e \frac{t_f k_r k_h f_{fud}}{D_{eq} f_{cd}} \geq 0.07 \end{cases} \quad (6.7)$$

Where

k_{cc} is the confinement factor for columns taken as $k_{cc} = 1.5$ for square and rectangular columns

D_{eq} is defined in Equation 6.1

k_e is the effectiveness factor for the applied system for a rectangular or square column, as is shown in Equation 6.8.

$$k_e = 1 - \frac{(b - 2\,r_c)^2 + (h - 2\,r_c)^2}{3bh} \quad (6.8)$$

k_r is a factor to account for the corner radius given in Equation 4.4, reproduced again below for clarity.

$$k_r = \begin{cases} \left(\frac{r_c}{50}\right)\left(2 - \frac{r_c}{50}\right) & \text{for } r_c < 50 \text{ mm} \\ 1 & \text{for } r_c \geq 50 \text{ mm} \end{cases}$$

k_h is a factor applied for systems that are wrapped intermittently or helically. Again, for clarity the factor has been applied in Equation 6.7 differently to *BS EN 1992-1-1:2023*, Formulae J.9 and J.10 and should be taken as shown in Equation 6.9.

$$k_h = \begin{cases} \left(1-\frac{(s_f-b_f)}{2b}\right)\left(1-\frac{(s_f-b_f)}{2h}\right)\left(\frac{1}{1+(tan\beta_f)^2}\right) & \text{for helical or intermittently wrapped systems} \\ 1 & \text{Otherwise} \end{cases} \tag{6.9}$$

The provision for square or rectangular columns was again taken from fib Bulletin No. 90 (fib, 2019) and manipulated to achieve the desired consistency with other provisions in *BS EN 1992-1-1:2023*. The strength enhancement factor k_{cc} was again reduced in the code to a value of 1.5 from 3.3 in fib Bulletin No. 90 (fib, 2019) to reflect the fact that a characteristic value is needed for design, as described in the Background Document to FprEN 1992-1-1 (European Committee for Standardisation, 2023).

The code states that the provisions do not apply where $\frac{h}{b} > 2$. This is because the effectiveness of strengthening rectangular sections becomes less and less effective as this aspect ratio increases. The limit of 2 has been set in this code.

General constitutive model for concrete in compression confined with CFRP

Concrete in compression has a substantial increase in ultimate strain as a result of application of the CFRP strengthening within the limits stated in the standard. To reflect this enhancement, the code modified the compressive response of concrete as per Equation 6.10. This model was developed by Darby *et al.* (2011) and Binici (2008). It is a simplification to the model of Lam and Teng (2003a and 2003b) commonly used in other international standards (e.g. ACI 440.2R-17 (ACI, 2017)). The model describes the increase in the rate of expansion for a given increase in axial load using a bilinear approximation; it takes the strain $\varepsilon_{c2,c} = 0.00175$ as the point at which the rate of expansion in the concrete increases for a given increment of stress.

The model takes a conservative lower bound of the ultimate strain of concrete confined with CFRP, $\varepsilon_{cu,c} = 0.006$.

$$\sigma_c = \begin{cases} \frac{f_{cd}\varepsilon_c}{0.00175} & for\ 0 < \varepsilon_c < 0.00175 \\ f_{cd} + \Delta f_{cd}\frac{(\varepsilon_c - 0.00175)}{(\varepsilon_{cu,c} - 0.00175)} & for\ 0.00175 < \varepsilon_c \leq \varepsilon_{cu,c} \end{cases} \tag{6.10}$$

Where

Δf_{cd} is calculated from Equation 6.5 or Equation 6.7 (Formulae J.6, J.7, J.9 and J.10 in *BS EN 1992-1-1:2023*).

Figure 6.4 shows a comparison of the bilinear stress-strain model used in *BS EN 1992-1-1:2023* with the ultimate limit state (ULS) model provided in Section 8.1.2 of the standard (Formula 8.4), for the

Figure 6.4 Example stress-strain diagram for a concrete member confined by CFRP strengthening

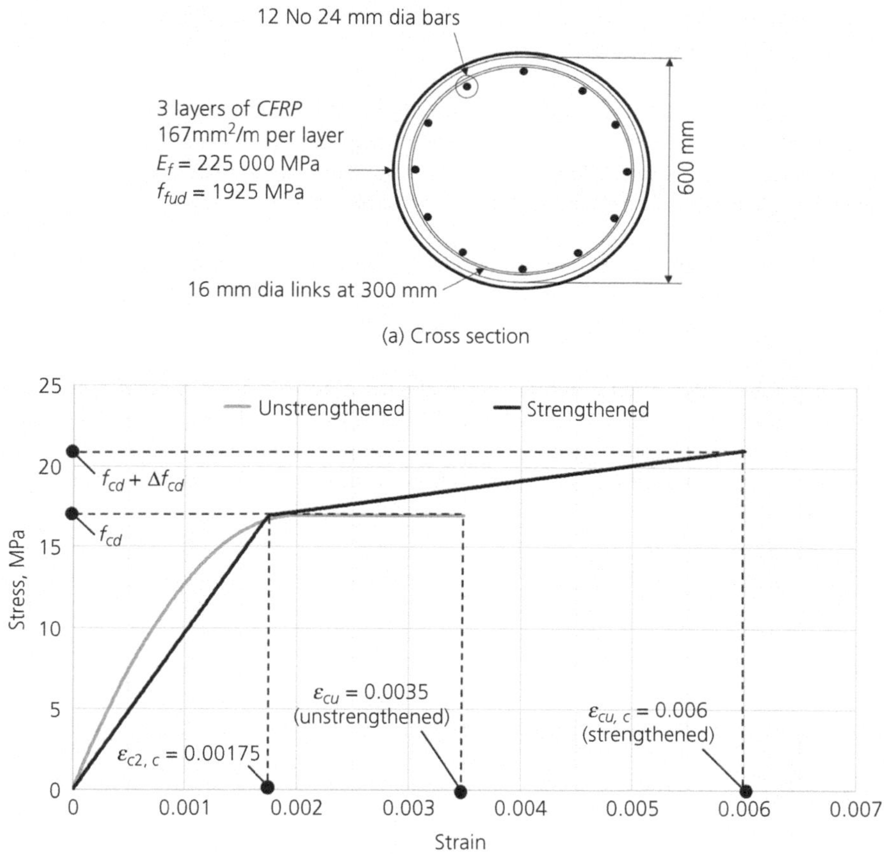

(b) Strengthened and unstrengthened stress-strain diagrams

column cross section documented in Figure 6.4(a). It shows an appreciable percentage increase in the ultimate resistance (24%) for the given cross section. The example assumes $f_{cd} = 17\text{MPa}$.

The CFRP wrapping system can fail at the overlap joint where there is insufficient overlap of the wrap at the point it is terminated. This will vary between systems and could not be codified; the supplier will need to be consulted to support in determining this value. This is also discussed in Chapter 9.

Eccentricity limit

The code is prescriptive in terms of the eccentricity limit that applies in determining actions for column analysis, which is stated in Equation 6.2. *BS EN 1992-1-1:2023* requires that this eccentricity on an applied axial load is taken into account as per Equation 6.11 unless second order effects and the effects of geometric imperfections, and imposed deformation, are taken into account.

$$e_{d,min} = max\begin{cases} \dfrac{h}{30} \\ 20\text{ mm} \end{cases} \tag{6.11}$$

Slenderness

As stated in *BS EN 1992-1-1:2023*, J.8.1.2(2), slender columns can be designed using the provisions in the standard provided that the slenderness falls within the limit in Equation 6.2. This invokes Section 7.4.1(3) and Annex O.

The ultimate compressive strain in confined columns is significantly higher than strain in unconfined columns, and the stiffness of the concrete in compression reduces significantly once the unconfined ultimate strain is exceeded. Annex J does not provide guidance on how to implement provisions in *BS EN 1992-1-1:2023* for columns strengthened with CFRP.

TR55 (Concrete Society, 2013) provides a method for determining when slenderness effects need to be applied. It also applies the nominal stiffness method (*BS EN 1992-1-1:2023*, O.7.3) to determine an additional moment to account for slenderness effects.

A parameter λ_{crit} for strengthened columns was introduced into TR55 (Concrete Society, 2013). It is similar to λ_{lim} as defined in *BS EN 1992-1-1:2023* but is modified to account for the reduction in concrete stiffness described above. For the purpose of applying *BS EN 1992-1-1:2023*, the TR55 provision may be used, if modified to suit the code provisions, as outlined in Equation 6.12.

$$\lambda_{crit} = \frac{\lambda_{lim}}{\dfrac{f_{cd,c}}{f_{cd}}\left(1+0.06\dfrac{\varepsilon_{h,rup}}{\varepsilon_{c2}}\right)} \qquad (6.12)$$

Where

λ_{lim} is calculated using Formula O.12 from *BS EN 1992-1-1:2023*

$$f_{cd,c} = f_{cd} + \Delta f_{cd}$$

Where

Δf_{cd} is calculated in accordance with *BS EN 1992-1-1:2023*, J.8.1.2

ε_{c2} is as defined in *BS EN 1992-1-1:2023*, Formula 8.4. $\varepsilon_{h,rup}$ is defined in Equation 6.13.

$\varepsilon_{h,rup}$ is the hoop rupture strain of the confining CFRP jacket as designed in Equation 6.13. The value of this strain is generally less than the ultimate tensile strain of the CFRP, the provision below has been determined using experimental results

$$\varepsilon_{h,rup} = \begin{cases} 0.6\dfrac{\varepsilon_{fud}}{\eta_f}, & \text{for circular columns} \\ \dfrac{\varepsilon_{fud}}{\eta_f}\left[0.46\left(\dfrac{2r_c}{h}\right)+0.14\right], & \text{for rectangular columns} \end{cases} \qquad (6.13)$$

BS EN 1992-1-1:2023 and TR55 (Concrete Society, 2013) use slightly different approaches for determining ε_{fud}, including partial factors and the use of mean or characteristic properties. However, when comparing results produced from the two methods, it can be seen that the values of ε_{fud}

are closely aligned, with only the exception of the value of η_f in the code. Reducing the value of the design tensile strength of the CFRP with the value $\eta_f = 0.7$ will have the effect of beneficially increasing the value of λ_{crit}. The reduction factor η_f only applies for long-term situations and should not be applied when determining λ_{crit}.

Where the column slenderness ratio defined in Equation 6.14 (and *BS EN 1992-1-1:2023*, 3.2.2 and Formula O.8) exceeds the value of λ_{lim}, methods shown in Section 7 and Annex O of the code need to be applied. TR55 (Concrete Society, 2013) used provisions that aligned with Section O.6 of *BS EN 1992-1-1:2023*.

$$\lambda = \frac{l_o}{i} \tag{6.14}$$

Based on guidance in TR55 (Concrete Society, 2013), the provisions in the nominal stiffness method should be used to determine the additional moment arising from slenderness effects.

Axial and bending (NM) interaction

When producing an NM interaction diagram, Equation 6.2 needs to be implemented. In this equation, the minimum value of eccentricity is provided at which point the beneficial effects of confinement need to be discounted because significant tension forces begin to prevail in the concrete, and the ability of the confining forces to be generated in the CFRP wrapping system can no longer be relied upon. Combining all relevant criteria and rewriting Equation 6.2, Table 6.1 provides the relevant formulations that may be used to construct an NM interaction diagram from first principles for a column strengthened with CFRP.

A method for developing a NM interaction diagram is presented in Figure 6.5. This is a strain-based method in which the strain at the outermost compression fibre is set to the ultimate compression strain. The ultimate compression strain is taken as either

- $\varepsilon_{cu,c} = 0.0006$, where the eccentricity satisfies the condition $e \leq 0.2D_{eq}$, or
- $\varepsilon_{cu} = 0.0035$ otherwise.

The method then reduces the strain value on the opposite surface step by step, generating successive axial force–moment pairs from the equilibrium of internal forces. The process continues until the resultant axial force reduces to zero, capturing the transition from compression-dominated to bending-dominated behaviour.

To obtain the full interaction diagram, results from both confined and unconfined concrete stress–strain models should be assembled. The tension limb of the curve can be simplified by considering the tension only resistance with a linear transition from pure bending to pure tension.

The examples produced in this chapter illustrate the impact the change in the stress-strain constitutive model has on the NM interaction diagram. TR55 (Concrete Society, 2013) contains some good guidance for reducing the number of calculation points on the curve, ensuring that key points are captured and simplifying this procedure for hand calculation. The same principles stated there are applicable to this method.

Columns confined and strengthened in flexure

Columns can be strengthened in flexure as well as providing confinement reinforcement, using independent strengthening systems applied to the same member. The flow chart outlined in

Table 6.1 Relevant equations for compiling NM interaction diagram

$\frac{M}{N} \leq 0.2D_{eq}$: Confined Concrete Model Applies	
Term	Definition
$\varepsilon_{c2,c}$	$\varepsilon_{c2,c} = 0.00175$
$\varepsilon_{cu,c}$	$\varepsilon_{cu,c} = 0.006$
σ_c	$\sigma_c = \begin{cases} \frac{f_{cd}\varepsilon_c}{\varepsilon_{c2,c}} & \text{for } 0 \leq \varepsilon_c \leq \varepsilon_{c2,c} \\ f_{cd} + \Delta f_{cd} \frac{(\varepsilon_c - \varepsilon_{c2,c})}{(\varepsilon_{cu,c} - \varepsilon_{c2,c})} & \text{for } \varepsilon_{c2,c} < \varepsilon_c \leq \varepsilon_{cu,c} \end{cases}$
Δf_{cd}	Circular Columns $\Delta f_{cd} = \begin{cases} 0 & \text{for } \frac{t_f k_h f_{fud}}{D f_{cd}} < 0.07 \\ k_{cc} \frac{t_f}{D} k_h f_{fud} & \text{for } \frac{t_f k_h f_{fud}}{D f_{cd}} \geq 0.07 \end{cases}$ Square or Rectangular Columns $\Delta f_{cd} = \begin{cases} 0 & \text{for } \left(\frac{b}{h}\right)^2 k_e \frac{t_f k_r k_h f_{fud}}{D_{eq} f_{cd}} < 0.07 \\ k_{cc} \left(\frac{b}{h}\right)^2 k_e \left(\frac{t_f}{D_{eq}}\right) k_r k_h f_{fud} & \text{for } \left(\frac{b}{h}\right)^2 k_e \frac{t_f k_r k_h f_{fud}}{D_{eq} f_{cd}} \geq 0.07 \end{cases}$
$\frac{M}{N} > 0.2D_{eq}$: Unconfined Concrete Model Applies	
ε_{c2}	$\varepsilon_{c2} = 0.002$
ε_{cu}	$\varepsilon_{cu} = 0.0035$
σ_c	$\sigma_c = \begin{cases} f_{cd}\left[1 - \left(1 - \frac{\varepsilon_c}{\varepsilon_{c2}}\right)^2\right] & \text{for } 0 \leq \varepsilon_c \leq \varepsilon_{c2} \\ f_{cd} & \text{for } \varepsilon_{c2} < \varepsilon_c \leq \varepsilon_{cu} \end{cases}$
Δf_{cd}	$\Delta f_{cd} = 0$

Figure 6.5 will generally apply for developing a NM interaction curve. The flexural strengthening would generally be placed between the concrete surface and the confining CFRP layer. When undertaking this combined strengthening, CFRP should not be used in compression.

When considering strengthening in flexure and for confinement, care will need to be taken with bond checks for the flexural strengthening system. While one might intuitively expect that anchorage and intermediate crack debonding resistances would benefit from the presence of confining

Figure 6.5 Flow chart for developing column curve using provisions of *BS EN 1992-1-1:2023*

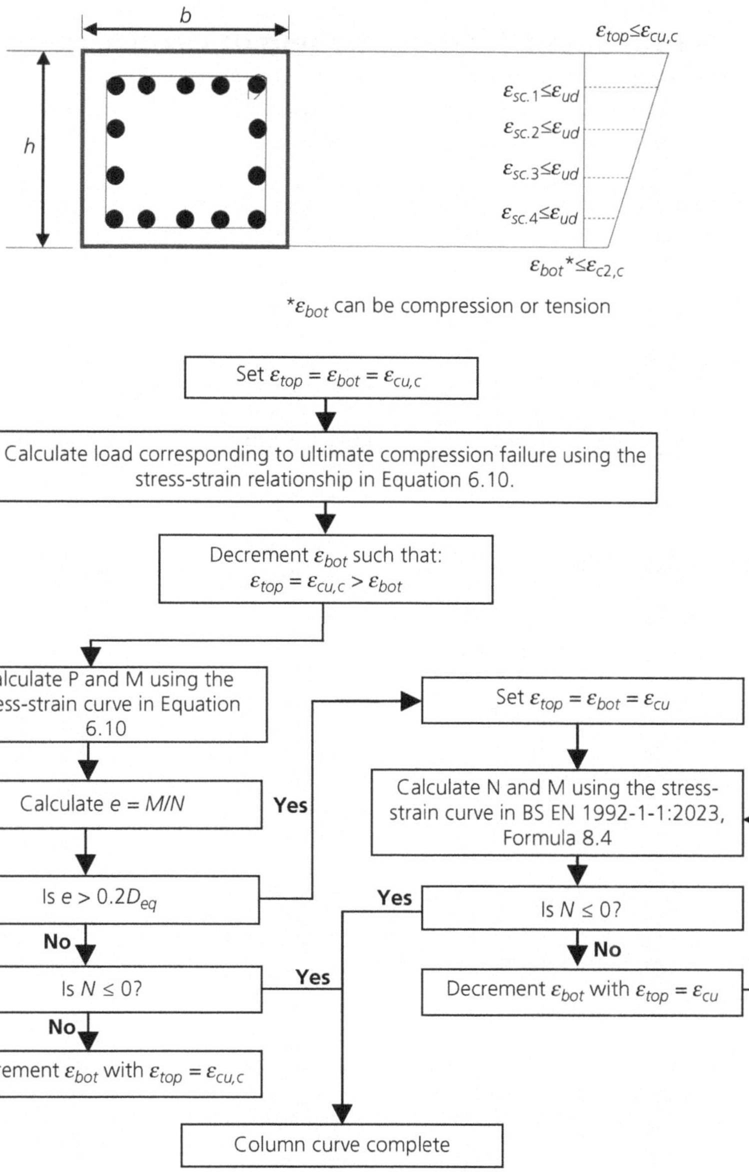

CFRP, there is no code model for this. One would also intuitively expect that the confinement can act as the shear reinforcement necessary to resist failure from end cover separation and shear induced cover separation. However, *BS EN 1992-1-1:2023* does not provide a method to superimpose effects from the addition of shear into the confinement CFRP to check that overstress of the CFRP does not occur. In these instances, it would be prudent to ensure that adequate anchorage is provided for the flexural strengthening, intermediate crack debonding checks pass, and that shear strengthening to resist end cover and shear induced separation is provided, additional to the confining CFRP, if needed.

Example Confinement 1: Strengthening of circular column with continuous wrapped system

Calculate the capacity of a circular column, with cross-sectional dimensions given in Figure 6.6, for EBR FRP strengthening using wet lay-up CF sheets.

Loading details

Applied axial force

$$N_{Ed} = 8000 \text{ kN}$$

Applied bending moment

$$e_{d,min} = max\left(\frac{h}{30};20 \text{ mm}\right) = max\left(\frac{600}{30};20 \text{ mm}\right) = 20 \text{ mm}$$

$$M_{Ed,min} = 160 \text{ kNm}$$

Reinforcement details

- Tensile reinforcement: 8 No 20 mm-diameter bars
- Concrete cover to main bars: 40 mm

Effective depths

- Layer 1: $d_{sc.1} = 69.03$ mm $A_{sc.1} = 628$ mm^2
- Layer 2: $d_{sc.2} = 204.32$ mm $A_{sc.2} = 628$ mm^2
- Layer 3: $d_{sc.3} = 395.67$ mm $A_{sc.3} = 628$ mm^2
- Layer 4: $d_{sc.4} = 530.97$ mm $A_{sc.4} = 628$ mm^2

Figure 6.6 Example Confinement 1 circular column with continuous wrapped system strengthening – all dimensions are in mm unless stated otherwise

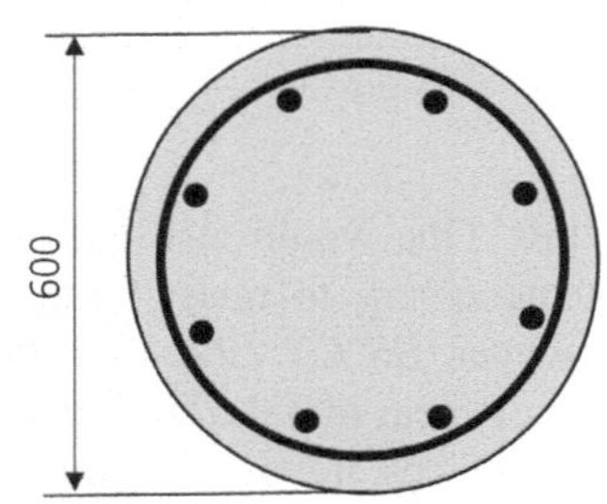

Materials

Concrete

Characteristic compressive strength: $f_{ck} = 45$ MPa

Partial safety factor (*BS EN 1992-1-1:2023*, Table 4.3): $\gamma_c = 1.5$

$$\eta_{cc} = \left(\frac{f_{ck,ref}}{f_{ck}}\right)^{\frac{1}{3}} = \left(\frac{40}{45}\right)^{\frac{1}{3}} = 0.96$$

Take $k_{tc} = 0.85$ (*BS EN 1992-1-1:2023*, 5.1.6 (1) NOTE)

Design compressive strength (*BS EN 1992-1-1:2023*, Formula 5.3)

$$f_{cd} = \frac{0.96 \times 0.85 \times 45}{1.5} = 24.5 \text{ MPa}$$

Steel reinforcement

Characteristic tensile strength: $f_{yk} = 500$ MPa

Modulus of elasticity: $E_s = 200\,000$ MPa

Partial safety factor (*BS EN 1992-1-1:2023*, Table 4.3), $\gamma_s = 1.15$

Design tensile strength (*BS EN 1992-1-1:2023*, Formula 5.11), $f_{yd} = \frac{f_{yk}}{\gamma_s} = \frac{500}{1.15} = 435$ MPa

FRP strengthening properties

Characteristic short-term tensile strength: $f_{fuk} = 2400$ MPa

Mean modulus of elasticity: $E_f = 225\,000$ MPa

Partial safety factor (*BS EN 1992-1-1:2023*, Table J.1): $\gamma_f = 1.4$

Reduction factor (*BS EN 1992-1-1:2023*, J.5.3 (2)): $\eta_f = 0.7$

Design tensile strength (*BS EN 1992-1-1:2023*, Formula J.2)

$$f_{fud} = \frac{\eta_f f_{fuk}}{\gamma_f} = \frac{0.7 \times 2400}{1.4} = 1200 \text{ MPa}$$

The cross-sectional area of the fibres per metre of a single layer of CF sheet given by the supplier is 478 mm²/m. Therefore, for a metre width of a CF sheet $A_{f,1} = 478$ mm² per single layer.

A maximum of ten CFRP layers can be applied in accordance with *BS EN 1992-1-1:2023*, J.12.4 (1).

Check existing capacity

Maximum axial capacity without bending moment

$$N_{Rd,pre} = 7963 \text{ kN} < N_{Ed} = 8000 \text{ kN}$$

Therefore for $N_{Ed} = 8000\,\text{kN}$

$$M_{Rd,pre} = 0 \text{ kNm} < M_{Ed,min} = 160 \text{ kNm}$$

Therefore, strengthening is required.

Check strengthened capacity with continuous CFRP wrapping

Check if $\frac{t_f k_h f_{fud}}{D f_{cd}} \geq 0.07$

For continuous wrapping: $k_h = 1.0$

Minimum two layers required for achieving confinement

$$t_f = n_f^{k_f} \frac{A_{f,1}}{b_f} = 2^1 \times 0.478 = 0.956 \text{ mm}$$

$$\frac{t_f k_h f_{fud}}{D f_{cd}} = \frac{0.956 \times 1.0 \times 1200}{600 \times 24.5} = 0.078 > 0.07 \text{ therefore there will be confining effects}$$

For circular columns, the following applies

$k_{cc} = 2.5$

$$\Delta f_{cd} = k_{cc}\left(\frac{t_f}{D}\right) k_h f_{fud} = 2.5 \times \left(\frac{0.956}{600}\right) \times 1.0 \times 1200 = 4.78 \text{ MPa}$$

The concrete stress-strain curves for the confined and unconfined concrete are shown in Figure 6.7 (a). The corresponding NM interaction curve Figure 6.7 (b) is constructed following the procedure outlined in Table 6.1 and using the iterative method described in Annex A. The bending moment capacity under the applied axial force is

$$M_{Rd} = 182.07 \text{ kNm} > M_{Ed,min} = 160 \text{ kNm} - \text{check is ok.}$$

Figure 6.7 Example Confinement 1 concrete: (a) concrete stress-strain curves, (b) NM interaction curve for the strengthened section

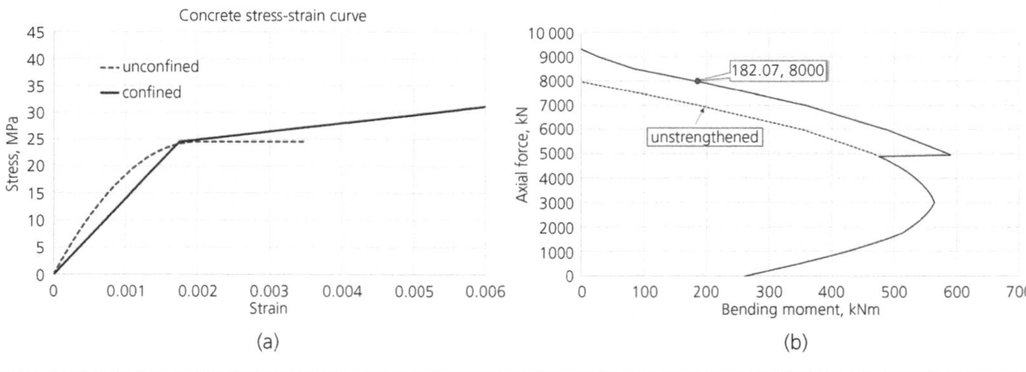

Example Confinement 2: Strengthening of square column with continuous wrapped system

Calculate the capacity of a square column, with cross-sectional dimensions given in Figure 6.8, for EBR FRP strengthening using wet lay-up CF sheets.

Loading details

Applied axial force

$$N_{Ed} = 10\,000 \text{ kN}$$

Applied bending moment

$$e_{d,min} = max(h/30; 20 \text{ mm}) = max(700/30; 20 \text{ mm}) = 23.3 \text{ mm}$$

$$M_{Ed,min} = 233 \text{ kNm}$$

Reinforcement details

- Tensile reinforcement: 12 No 25 mm-diameter bars
- Shear links: 10mm-diameter stirrups at 300 mm spacing
- Concrete cover to links: 50 mm

Effective depths

- Layer 1: $d_{sc.1} = 627.5\,\text{mm}$ $A_{sc.1} = 1963\,\text{mm}^2$
- Layer 2: $d_{sc.2} = 442.5\,\text{mm}$ $A_{sc.2} = 982\,\text{mm}^2$
- Layer 3: $d_{sc.3} = 257.5\,\text{mm}$ $A_{sc.3} = 982\,\text{mm}^2$
- Layer 4: $d_{sc.4} = 72.5.5\,\text{mm}$ $A_{sc.4} = 1963\,\text{mm}^2$

Figure 6.8 Example Confinement 2 rectangular column with continuous wrapped system strengthening – all dimensions are in mm unless stated otherwise

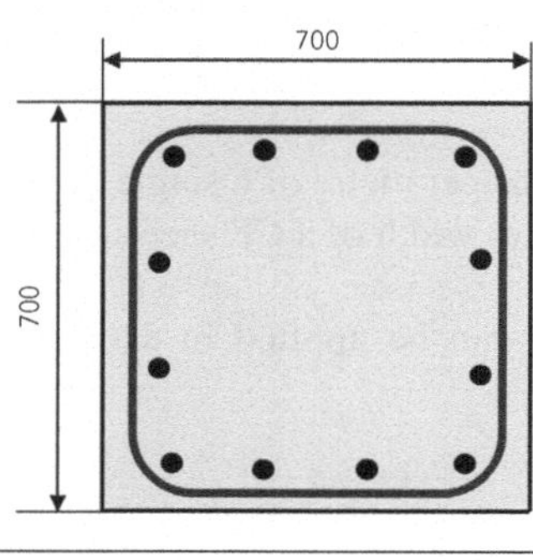

Materials

Concrete

Characteristic compressive strength: $f_{ck} = 30$ MPa

Partial safety factor (*BS EN 1992-1-1:2023*, Table 4.3): $\gamma_c = 1.5$

$$\eta_{cc} = \left(\frac{f_{ck,ref}}{f_{ck}}\right)^{\frac{1}{3}} = \left(\frac{40}{30}\right)^{\frac{1}{3}} = 1.1 > 1.0, \quad \text{therefore } \eta_{cc} = 1.0$$

Take $k_{tc} = 0.85$ (*BS EN 1992-1-1:2023*, 5.1.6 (1) NOTE)

Design compressive strength (*BS EN 1992-1-1:2023*, Formula 5.3)

$$f_{cd} = \frac{1 \times 0.85 \times 30}{1.5} = 17 \text{ MPa}$$

Steel reinforcement

Characteristic tensile strength: $f_{yk} = 400$ MPa

Modulus of elasticity: $E_s = 200\,000$ MPa

Partial safety factor (*BS EN 1992-1-1:2023*, Table 4.3): $\gamma_s = 1.15$

Design tensile strength (*BS EN 1992-1-1:2023*, Formula 5.11): $f_{yd} = \dfrac{f_{yk}}{\gamma_s} = \dfrac{400}{1.15} = 348$ MPa

FRP strengthening properties

Characteristic short-term tensile strength: $f_{fuk} = 2400$ MPa

Mean modulus of elasticity: $E_f = 225\,000$ MPa

Partial safety factor (*BS EN 1992-1-1:2023*, Table J.1): $\gamma_f = 1.4$

Reduction factor (*BS EN 1992-1-1:2023*, J.5.3 (2)): $\eta_f = 0.7$

Design tensile strength (*BS EN 1992-1-1:2023*, Formula J.2):

$$f_{fud} = \frac{\eta_f f_{fuk}}{\gamma_f} = \frac{0.7 \times 2400}{1.4} = 1200 \text{ MPa}$$

The cross-sectional area of the fibres per metre of a single layer of CF sheet given by the supplier is 478 mm^2/m. Therefore, for a metre width of a CF sheet $A_{f,1} = 478$ mm^2 per single layer

A maximum of ten CFRP layers can be applied in accordance with *BS EN 1992-1-1:2023*, J.12.4 (1).

Check existing capacity

Calculate bending moment capacity for $N_{Ed} = 10\,000$ kN

$$M_{Rd,pre} = 79 \text{ kNm} < M_{Ed,min} = 233 \text{ kNm}$$

Therefore, strengthening is required.

Check strengthened capacity with continuous CFRP wrapping

$$D_{eq} = \frac{2\,bh}{b+h} = \frac{2\times700\times700}{700+700} = 700 \text{ mm}$$

Assume $r_c = 30$ mm, therefore

$$k_r = 1\left(\frac{r_c}{50}\right)\left(2-\frac{r_c}{50}\right) = 1\times\left(\frac{30}{50}\right)\left(2-\frac{30}{50}\right) = 0.84$$

$$k_e = 1-\frac{(b-2r_c)^2+(h-2r_c)^2}{3bh} = 1-\frac{(700-2\times30)^2+(700-2\times30)^2}{3\times700\times700} = 0.44$$

$$k_h = 1.0$$

Check if $\left(\frac{b}{h}\right)^2 k_e \frac{t_f k_r k_h f_{fud}}{D_{eq} f_{cd}} \geq 0.07$

Five layers will be sufficient to achieve the required strengthened capacity.

$$t_f = n_f^{k_f}\frac{A_{f,1}}{b_f} = 5^{0.85}\times0.478 = 1.877 \text{ mm}$$

$$\left(\frac{b}{h}\right)^2 k_e \frac{t_f k_r k_h f_{fud}}{D_{eq} f_{cd}} = 1\times\ 0.44\times\frac{1.877\times0.84\times1.0\times1200}{700\times17} = 0.0704 > 0.07$$

– check is ok.

For square columns, the following applies

$$k_{cc} = 1.5$$

$$\Delta f_{cd} = k_{cc}\left(\frac{b}{h}\right)^2 k_e\left(\frac{t_f}{D_{eq}}\right)k_r k_h f_{fud} = 1.5\times1\times0.44\times\left(\frac{1.877}{700}\right)\times0.84\times1.0\times1\,200 = 1.80 \text{ MPa}$$

The NM interaction curve, shown in Figure 6.9 (b), is constructed following the procedure outlined in Table 6.1 and using the iterative method described in Annex A for the concrete stress-strain curves shown in Figure 6.9 (a). The bending moment capacity under the applied axial force is

$$M_{Rd} = 270.80 \text{ kNm} > M_{Ed,min} = 233 \text{ kNm} \text{ – check is ok.}$$

Figure 6.9 Example Confinement 2: (a) concrete stress-strain curves, (b) NM interaction curve for the strengthened section

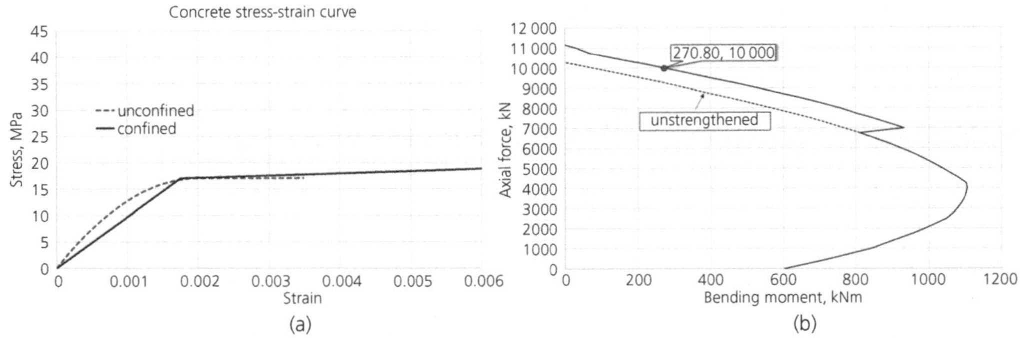

Example Confinement 3: Strengthening of circular column with discrete CFRP strips

For the circular column in Example Confinement 1, calculate the strengthened capacity using discrete CFRP strips assuming the following.

CFRP strips properties

Characteristic short-term tensile strength: $f_{fuk} = 2900$ MPa (5% fractile)

Mean modulus of elasticity: $E_f = 170\,000$ MPa

Check strengthened capacity

Partial safety factor (*BS EN 1992-1-1:2023*, Table J.1), $\gamma_f = 1.3$

Reduction factor (*BS EN 1992-1-1:2023*, J.5.3 (2)), $\eta_f = 0.7$

Design tensile strength (*BS EN 1992-1-1:2023*, Formula J.2)

$$f_{fud} = \frac{\eta_f f_{fuk}}{\gamma_f} = 0.7 \times \frac{2900}{1.3} = 1562 \text{ MPa}$$

$$t_f = 1.2 \text{ mm}$$

$$b_f = 200 \text{ mm}$$

$$s_f = 300 \text{ mm}$$

$$\beta_f = 0$$

For helical and/or intermittent wrapping

$$k_h = \left(1 - \frac{\left(s_f - b_f\right)}{2D}\right)^2 \times \left(\frac{1}{1+\left(tan\beta_f\right)^2}\right) = \left(1 - \frac{\left(300-200\right)}{2\times 600}\right)^2 \times \left(\frac{1}{1+\left(tan0\right)^2}\right) = 0.84$$

Check if $\frac{t_f k_h f_{fud}}{Df_{cd}} \geq 0.07$

$$\frac{t_f f_{fud}}{Df_{cd}} = \frac{1.2\times 0.84\times 1562}{600\times 24.5} = 0.107 > 0.07$$

Therefore, the thickness is sufficient for achieving confining effect.

$$\Delta f_{cd} = k_{cc}\left(\frac{t_f}{D}\right)k_h f_{fud} = 2.5\times\left(\frac{1.2}{600}\right)\times 0.84\times 1562 = 6.56 \text{ MPa}$$

From the NM interaction curve in Figure 6.10 (b), constructed based on the concrete stress-strain curves shown in Figure 6.10 (a), following the procedure outlined in Table 6.1 and using the iterative method described in Annex A, the bending moment capacity is adequate under the applied axial force.

$$M_{Rd} = 238.22 \text{ kNm} > M_{Ed,min} = 160 \text{ kNm}$$

Figure 6.10 Example Confinement 3: (a) concrete stress-strain curves, (b) NM interaction curve for the strengthened section

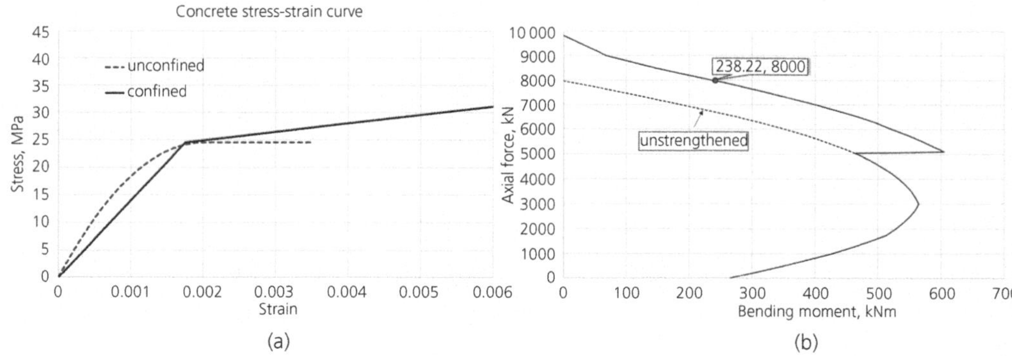

Kaloyana Kostova and Craig Giaccio
ISBN 978-1-83549-935-1
https://doi.org/10.1108/978-1-83549-934-420251007

Chapter 7
Serviceability

Introduction

BS EN 1992-1-1:2023 (BSI, 2023b), J.9 is a very brief part of the standard. It does not exclude any of the provisions within the main standard, suggesting that all checks in *BS EN 1992-1-1:2023* apply when strengthening with CFRP. Serviceability limit state (SLS) checks may be needed where ultimate limit state (ULS) strengthening is required. The following points are specifically referred to in the code.

- When determining deflections in a member which comprises three different materials, the section properties can be transformed using reinforced concrete beam theory. In doing so, the slip between the CFRP and concrete substrate will be ignored, as it is when transforming concrete to steel (or vice versa).
- As was discussed for ULS considerations (in Chapters 3–5), the section properties change when adding CFRP. Superposition may be used to deal with summation of deflections, pre- and post-strengthening, to cater for the change in section properties.
- Superposition can also be used to determine stress in the concrete and reinforcement/ prestressing steel.
- Stresses must be limited to retain bond durability. As is outlined in this chapter, section properties calculated with long-term effects need the long-term modulus of elasticity of the concrete to be used. This is defined in *BS EN 1992-1-1:2023*, 9.1(4) and is illustrated throughout the text and examples in this chapter.

CFRP cannot sustain significant amounts of compression. It should not be used in compression areas, and care must be taken in members where creep may be significant (e.g. prestressed concrete beams) that this creep does not impart significant compression onto the CFRP strengthening system.

Stress limits

The stress limit given in Equation 7.1 (*BS EN 1992-1-1:2023*, Formula J.24) under serviceability conditions is included to prevent damage or significant creep of the concrete and adhesive, and also to avoid stress rupture of the CFRP strips or sheets.

The code requires that the stress in the CFRP does not exceed a value where the reinforcement stress is greater than 80% of the yield stress. This ensures that the reinforcement cannot yield, thereby removing significant strain gradients in the CFRP at SLS, and also ruling out significant changes in strain gradient in the CFRP post-strengthening, and associated bond damage.

$$\sigma_f \leq 0.8 f_{yk} \frac{E_f}{E_s} \tag{7.1}$$

The limit in Equation 7.1 is to be considered in addition to the stress limits for concrete, reinforcement and prestressing steel stated in Annex I and Section 9 of *BS EN 1992-1-1:2023*. As discussed above, to determine stresses in the section at SLS, the pre-strengthened and strengthened state need to be considered.

To undertake stress checks of a member strengthened with CFRP, the following approach may be applied. Consider that the structure in Figure 7.1 is subjected to an applied ULS moment of $M_{Ed,SLS}$, which can be broken down into two components as defined in Equation 7.2.

$$M_{Ed,SLS} = M_{Ed,str} + M_{post} \tag{7.2}$$

Where

$M_{Ed,str}$ is the moment at strengthening

M_{post} is the additional moment applied after strengthening.

The superposition of stresses that arise from the applied moments pre- and post-strengthening is illustrated in Figure 7.2. This figure implies a linear stress distribution in the compression area. This may be used where stress levels are relatively low. If higher stress levels prevail, it would be more accurate to use the general stress-strain relationships given in *BS EN 1992-1-1:2023*, 5.1.6(3).

Figure 7.1 Arbitrary section for flexural strengthening

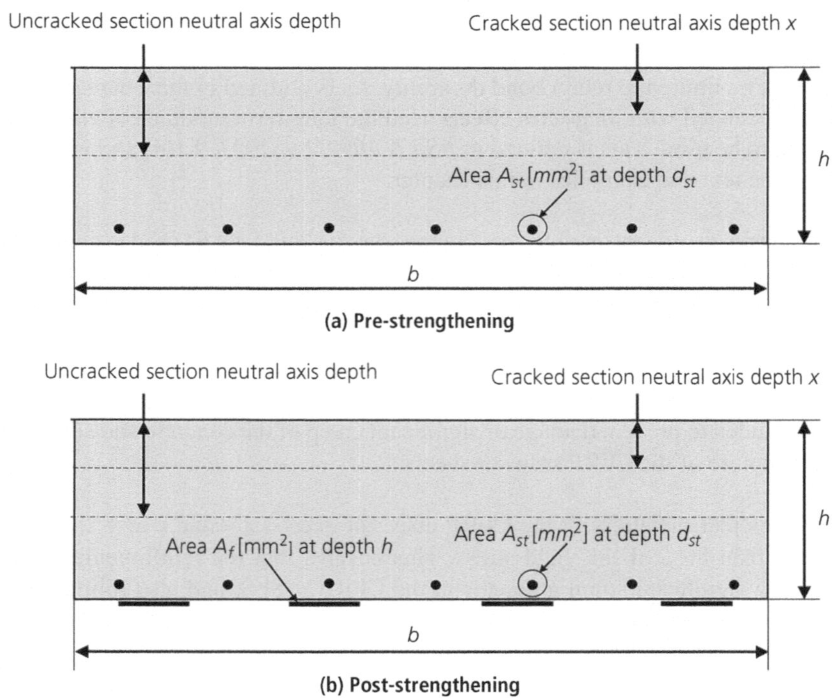

Figure 7.1 Continued

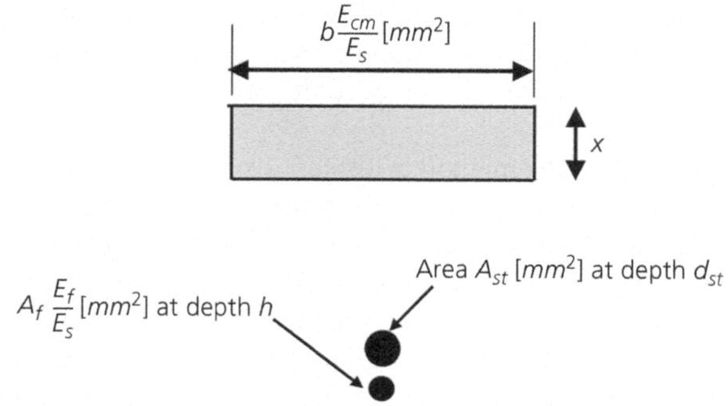

(c) Post-strengthening, cracked and transformed to steel

Figure 7.2 Stress superposition at serviceability limit state

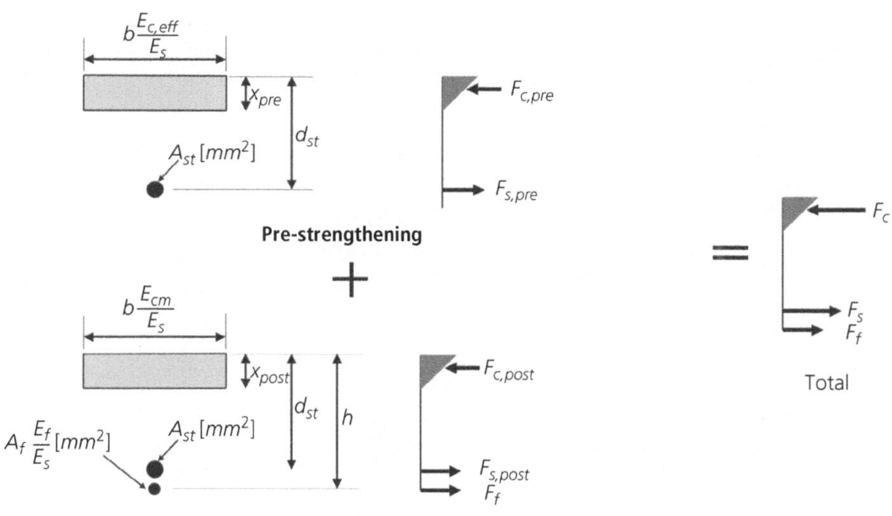

For convenience, and given this is quite cumbersome to use, Annex E provides expressions for the net compression force and lever arm using this method that can be obtained for a rectangular compression region using BS EN 1992-1-1(2023) Formula 5.1.6(3) which can directly be applied for SLS calculations. Methods presented in this book assume a linear strain distribution.

With linear stress distributions, cracked section analysis similar to that presented in Hendy *et al.* (2007) may be used for serviceability checks. To apply this method, the section needs to be transformed into a single material. Considering, for example, the system shown in Figure 7.1, the cross section can be transformed into steel by reducing the concrete width using the modular ratio $\frac{E_{cm}}{E_s}$. Similarly, the area of the CFRP is modified to an equivalent steel area by applying the modular ratio $\frac{E_f}{E_s}$.

The modular ratio shown in Figure 7.1 is the short-term ratio based on E_{cm}. If long-term actions were to be considered, the ratio would be modified to use $E_{c,eff}$ as set out in *BS EN 1992-1-1:2023*, 9.1(4). In the remainder of this section, it is assumed that pre-strengthening load effects are longterm and need the effective concrete modulus of elasticity to be used, whereas those after strengthening are shortterm (transient) and only need the mean modulus of elasticity to be used.

To calculate stresses in the section, the method in Table 7.1 may be used. In addition to the assumptions already stated above, this method also neglects the contribution of any compression reinforcement and assumes that the section is cracked pre- and post-strengthening. As stated in *BS EN 1992-1-1:2023*, Clause 9.1(2), if the value of stress calculated under the characteristic combination of actions does not exceed the value of stress at which cracking is expected to occur, $f_{ct,eff}$, then gross section properties may be used. Note that the method for determination of $f_{ct,eff}$ is discussed in *BS EN 1992-1-1:2023*, 9.1(3). The necessary checks to assess this condition are shown in the examples that follow.

Table 7.1 Calculation of SLS stresses pre- and post-strengthening

Pre-strengthening	Post-strengthening
Summing forces in the section $F_c = F_s$ When considering a linear compressive stress distribution, this yields $\frac{x_{pre}}{2}\varepsilon_{c,pre}E_{c,eff}b = A_{st}E_s\varepsilon_s$ Where the absolute value of the steel strain is given by $\varepsilon_{st} = \varepsilon_{c,pre}\left(\frac{d_{st} - x_{pre}}{x_{pre}}\right)$ Therefore $E_{c,eff}b\frac{x_{pre}}{2} = A_{st}E_s\left(\frac{d_{st} - x_{pre}}{x_{pre}}\right)$ Which yields $x_{pre} = \frac{-A_{st}E_s + \sqrt{(A_{st}E_s)^2 + 2bA_{st}E_sE_{c,eff}d_{st}}}{bE_{c,eff}}$	Summing forces in the section $F_c = F_s + F_f$ By lumping the reinforcement and CFRP into a single equivalent layer, as was done in Chapter 5, the following may be used for a beam with a single layer of reinforcement $A_{equiv} = A_{st} + A_f\frac{E_f}{E_s}$ and $d_{equiv} = \frac{d_fE_fA_f + d_{st}E_sA_{st}}{E_fA_f + E_sA_{st}}$ This may be used to evaluate $F_s + F_f$ as follows. $F_s + F_f = \varepsilon_{equiv}A_{equiv}E_s$ Where ε_{equiv} is the strain at the level d_{equiv}, the absolute value of which can be determined using $\varepsilon_{equiv} = \varepsilon_c\left(\frac{d_{equiv} - x}{x}\right)$ Therefore $\frac{bx}{2}E_{cm} = \left(\frac{d_{equiv} - x}{x}\right)A_{equiv}E_s$

Table 7.1 Continued

To calculate the second moment of area, concrete is also transformed into an equivalent steel area, as shown in Figure 7.2. Summing second moments of area about the neutral axis, the cracked second moment of area (moment of inertia) post-strengthening, $I_{cr,pre}$, is calculated using	Which yields $x = \frac{-A_{equiv}E_s + \sqrt{\left(A_{equiv}E_s\right)^2 + 2bA_{equiv}E_sE_{cm}d_{equiv}}}{bE_{cm}}$ Transforming the cross section to steel, and summing second moments of area about the neutral axis, the cracked second moment of area (moment of inertia), post-strengthening, $I_{cr,post}$, is calculated using
$I_{cr,pre} = \frac{bx_{pre}^{\;3}}{3}\frac{E_{c,eff}}{E_s} + A_{st}\left(d_{st} - x_{pre}\right)^2$	$I_{cr,post} = \frac{bx^3}{3}\frac{E_{cm}}{E_s} + A_{st}\left(d_{st} - x\right)^2 + A_{f,eff}\left(d_f - x\right)^2$ Where $A_{f,eff} = A_f\frac{E_f}{E_s}$

Superposition

The stress in the CFRP may be determined as shown in Equation 7.3.

$$\sigma_f = \frac{M_{post}\left(d_f - x\right)}{I_{cr,post}}\frac{E_f}{E_s} \tag{7.3}$$

Where

$$M_{post} = M_{Ed,SLS} - M_{Ed,str}$$

The concrete stress may be determined as shown in Equation 7.4.

$$\sigma_c = \sigma_{c,pre} + \sigma_{c,post} = \frac{M_{Ed,str}x_{pre}}{I_{cr}}\frac{E_{c,eff}}{E_s} + \frac{M_{post}x}{I_{cr,post}}\frac{E_{cm}}{E_s} \tag{7.4}$$

The steel stress may be determined as shown in Equation 7.5.

$$\sigma_s = \sigma_{s,pre} + \sigma_{s,post} = \frac{M_{Ed,str}\left(d_{st} - x_{pre}\right)}{I_{cr,pre}} + \frac{M_{post}\left(d_{st} - x\right)}{I_{cr,post}} \tag{7.5}$$

Example EBR Serviceability 1: Stress checks

The reinforced concrete slab structure in Example Flexural Design 1 (see Chapter 5) is subjected to a sagging bending moment of 120 kNm/m under a characteristic combination of actions at SLS, with 50 kNm/m being from self-weight and superimposed dead load, and 70 kNm/m from live load.

It has been identified that the slab requires strengthening for ULS with an EBR system comprising CFRP strips at 250 mm centres with $t_f = 1.2$ mm $b_f = 80$ mm, $f_{fud} = 1562$ MPa, and $E_f = 170\,000$ MPa. Check all serviceability stresses in the slab after strengthening. Undertake the stress checks in accordance with Section J.9 of *BS EN 1992-1-1:2023*.

Assume that the strengthening is undertaken with no live load present. Also assume that cracking occurs under the characteristic combination of actions and ignore compression reinforcement in the calculation. Assume a value of $\varphi(t,t_f)=1.6$, where *BS EN 1992-1-1:2023* Annex B is not required for detailed calculation of the creep coefficient.

$$E_{cm} = 33\,282 \text{ MPa}$$

$$A_f = 384 \text{ mm}^2$$

As stated in *BS EN 1992-1-1:2023*, Formula 9.1, assuming $\sigma_c \le 0.4 f_{cm}$

$$E_{c,eff} = \frac{1.05 E_{cm}}{1+\varphi(t,t_0)} = \frac{1.05 \times 33\,282}{1+1.6} = 13\,441 \text{ MPa}$$

$$x_{pre} = \frac{-A_{st}E_s + \sqrt{(A_{st}E_s)^2 + 2bA_{st}E_sE_{c,eff}d_{st}}}{bE_{c,eff}}$$

$$= \frac{-2010 \times 200\,000 + \sqrt{(2010 \times 200\,000)^2 + 2 \times 1000 \times 2010 \times 200\,000 \times 13\,441 \times 192}}{1000 \times 13\,441}$$

$$= 81.4 \text{ mm}$$

$$I_{cr,pre} = \frac{bx_{pre}^3}{3}\frac{E_{c,eff}}{E_s} + A_{st}(d_{st} - x_{pre})^2 = \frac{1000 \times 81.4^3}{3} \times \frac{13\,441}{200\,000} + 2010 \times (192 - 81.4)^2 = 36.7 \times 10^6 \text{ mm}^4$$

$$A_{equiv} = A_{st} + A_f\frac{E_f}{E_s} = 2010 + 384 \times \frac{170\,000}{200\,000} = 2336 \text{ mm}^2$$

$$d_{equiv} = \frac{d_fE_fA_f + d_{st}E_sA_{st}}{E_fA_f + E_sA_{st}} = \frac{250 \times 170\,000 \times 384 + 192 \times 200\,000 \times 2010}{170\,000 \times 384 + 200\,000 \times 2010} = 200.1 \text{ mm}$$

$$x = \frac{-A_{equiv}E_s + \sqrt{(A_{equiv}E_s)^2 + 2bA_{equiv}E_sE_c\mathrm{d}_{equiv}}}{bE_c}$$

$$= \frac{-2336 \times 200\,000 + \sqrt{(2336 \times 200\,000)^2 + 2 \times 1000 \times 2336 \times 200\,000 \times 33\,282 \times 200.1}}{1000 \times 33\,282}$$

$$= 62.2 \text{ mm}$$

$$I_{cr,post} = \frac{bx^3}{3}\frac{E_c}{E_s} + A_{st}\left(d_{st} - x\right)^2 + A_f\frac{E_f}{E_s}\left(d_f - x\right)^2$$

$$= \frac{1000 \times 62.2^3}{3}\frac{33282}{200000} + 2010 \times \left(192 - 62.2\right)^2$$

$$+ 384 \times \frac{170000}{200000} \times \left(250 - 62.2\right)^2 = 58.7 \times 10^6 \text{ mm}^4$$

Stress limits are as follows.

$$\sigma_f \leq 0.8 f_{yk}\frac{E_f}{E_s} = 0.8 \times 500 \times \frac{170000}{200000} = 340 \text{ MPa}$$

$$\sigma_c \leq 0.6 f_{ck} = 21 \text{ MPa}$$

$$\sigma_s \leq 0.8 f_{yk} = 400 \text{ MPa}$$

SLS stresses in each of the materials (Equation 7.3, Equation 7.4 and Equation 7.5) are as follows.

$$\sigma_f = \frac{M_{post}\left(d_f - x\right)}{I_{cr,post}}\frac{E_f}{E_s} = \frac{70 \times 10^6 \times (192 - 62.2)}{58.7 \times 10^6} \times \frac{170000}{200000}$$

$$= 190.3 \text{ MPa} < 340 \text{ MPa} - \text{check is ok.}$$

$$\sigma_c = \sigma_{c,pre} + \sigma_{c,post} = \frac{M_{Ed,str}x_{pre}}{I_{cr}}\frac{E_{c,eff}}{E_s} + \frac{M_{post}x}{I_{cr,post}}\frac{E_c}{E_s}$$

$$= \frac{50 \times 10^6 \times 81.4}{36.7 \times 10^6} \times \frac{13441}{200000} + \frac{70 \times 10^6 \times 62.2}{58.7 \times 10^6} \times \frac{33282}{200000}$$

$$\sigma_c = 7.5 + 12.3 = 19.8 \text{ MPa} < 21 \text{ MPa} - \text{check is ok.}$$

$$\sigma_s = \sigma_{s,pre} + \sigma_{s,post} = \frac{M_{Ed,str}\left(d_{st} - x_{pre}\right)}{I_{cr,pre}} + \frac{M_{post}\left(d_{st} - x\right)}{I_{cr,st}}$$

$$= \frac{50 \times 10^6 \times \left(192 - 81.4\right)}{36.7 \times 10^6} + \frac{70 \times 10^6 \times \left(192 - 62.2\right)}{58.7 \times 10^6}$$

$$\sigma_s = 150.9 + 154.7 = 305.6 \text{ MPa} < 400 \text{ MPa} - \text{check is ok.}$$

Example EBR Serviceability 2: Concrete stress limits in a prestressed concrete slab

A post-tensioned slab has been inspected and tested for determining the residual prestress in selected strands. The arrangement of the tendons at midspan is shown in Figure 7.3. Each tendon consists of four strands initially prestressed at 115 kN. Several tendons have been found to be loose. As a result, only seven out of ten tendons in each layer have been assumed to be working. In addition, based on site measurements, the residual prestress in tested tendons has been found to be 75 kN.

Figure 7.3 Example post-tensioned slab original details – all dimensions are in mm unless stated otherwise

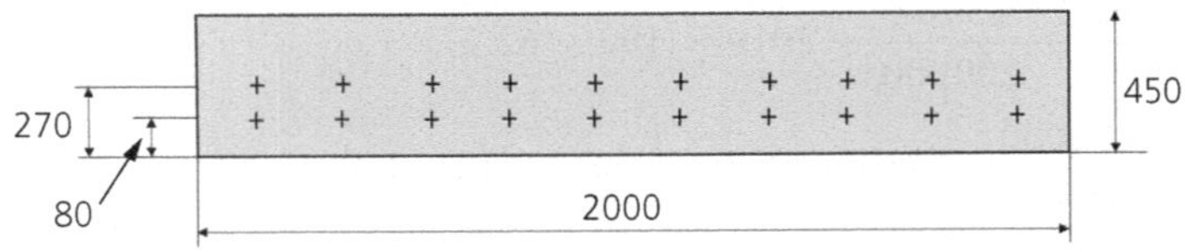

The ULS capacity of the slab for the reduced number of tendons and the reduced prestress has been found to be adequate. However, the SLS stress limit in tension has been found to be exceeded by more than 5%. The defined SLS stress limits are 12.5 MPa in compression and 2.65 MPa in tension.

Externally bonded strengthening with CF sheets has been considered to reduce the tensile stress in concrete at SLS. Calculate the stresses in the strengthened slab under the applied bending moment at SLS.

Applied load effects

Permanent maximum bending moment: 265 kNm

Live maximum bending moment: 450 kNm

Total bending moment at SLS: $265 + 450 = 715$ kNm

All strands originally stressed to 115 kN

Residual prestress per strand: 75 kN

Total prestressing force per tendon: $4 \times 75 = 300$ kN

Materials

CONCRETE

Characteristic compressive strength: $f_{ck} = 50$ MPa

Mean compressive strength (*BS EN 1992-1-1:2023*, Table 5.1): $f_{cm} = 58$ MPa

Modulus of elasticity (*BS EN 1992-1-1:2023*, Formula 5.1):

$$E_{cm} = k_E f_{cm}^{\frac{1}{3}} = 9500 \times 58^{\frac{1}{3}} = 36\,733 \text{ MPa}$$

Partial safety factor for SLS (*BS EN 1992-1-1:2023*, Table 4.3): $\gamma_c = 1.0$

$$\eta_{cc} = \left(\frac{f_{ck,ref}}{f_{ck}}\right)^{\frac{1}{3}} = \left(\frac{40}{50}\right)^{\frac{1}{3}} = 0.93 < 1.0$$

Take $k_{tc} = 0.85$ (*BS EN 1992-1-1:2023*, 5.1.6 (1) NOTE)

$$f_{cd} = \frac{0.93 \times 0.85 \times 50}{1.0} = 39.5 \text{ MPa}$$

FRP STRENGTHENING PROPERTIES

Characteristic short-term tensile strength

$$f_{fuk} = 4400 \text{ MPa}$$

Mean modulus of elasticity

$$E_f = 240\,000 \text{ MPa}$$

Partial safety factor (*BS EN 1992-1-1:2023*, Table J.1 for CF sheets)

$$\gamma_f = 1.4$$

Reduction factor (*BS EN 1992-1-1:2023*, J.5.3 (2))

$$\eta_f = 0.7$$

$$f_{fud} = \frac{0.7 \times 440}{1.4} = 2200 \text{ MPa}$$

Design ultimate strain (*BS EN 1992-1-1:2023*, Formula J.3)

$$\varepsilon_{fud} = \frac{f_{fud}}{E_f} = \frac{2200}{240\,000} = 0.00917 = 0.917\%$$

Width: $b_f = 1800$ mm

The cross-sectional area of the fibres per metre of a single layer of CF sheet given by the supplier is 167 mm²/m. Therefore, for a CF sheet with $b_f = 1800$ mm

$A_{f,1} = 167 \times 1.8 = 300.6$ mm² per single layer

A maximum of five CFRP layers can be applied in accordance with *BS EN 1992-1-1:2023*, J.12.4 (1).

Effective depth of CFRP reinforcement (taken at the extreme concrete fibre)

$$d_f = 450 \text{ mm}$$

Three layers of CF sheets will be required based on initial calculations. Therefore, in accordance with *BS EN 1992-1-1:2023*, J.5.2(4)

$$k_f = 1$$

Calculate the effective thickness (*BS EN 1992-1-1:2023*, Formula J.1)

$$t_f = n_f^{\,k_f} \frac{A_{f,1}}{b_f} = 3^1 \frac{300.6}{1800} = 0.501 \text{ mm}$$

Cross-sectional area: $A_f = 0.501 \times 1800 = 901.8\ \text{mm}^2$

Check SLS stresses in unstrengthened section

Calculate the total prestressing force

$$P = 2 \times 7 \times 300 = 4200\ \text{kN}$$

Calculate the prestressing moment about gross section centroid

$$M_p = \left[7 \times 300 \times (450 - 80 - 225) + 7 \times 300 \times (450 - 270 - 225)\right] \times 10^{-3} = 210\ \text{kNm}$$

Calculate the total stresses, shown in Figure 7.4.

Top surface

12.1 MPa < 12.5 MPa – check is ok.

Bottom surface

2.81 MPa > 2.65 MPa – check fails.

Calculate pre-strain in concrete at strengthening

The stresses in the section at strengthening are shown in Figure 7.5. The pre-strain at the concrete bottom surface is calculated accounting for long-term effects (refer to Example EBR serviceability 1 for detailed calculation) using long-term effective concrete modulus of elasticity $E_{c,eff} = 18367$ MPa.

Therefore

$$\varepsilon_0 = \frac{3.9}{18367} = 0.00021$$

Figure 7.4 Stress distribution at SLS – post-tensioned slab

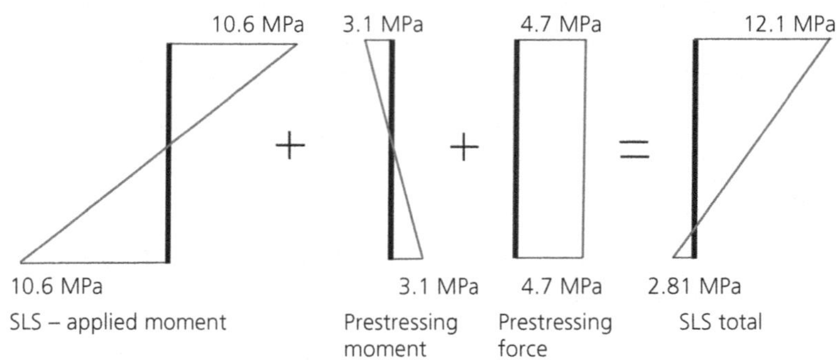

Figure 7.5 Stress distribution at strengthening – post-tensioned slab

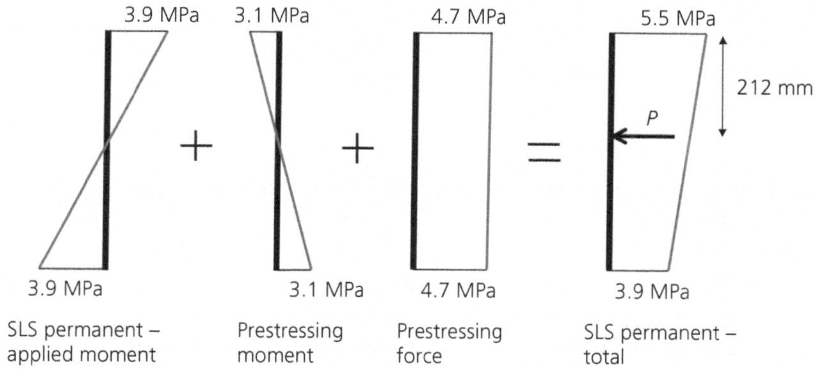

Figure 7.6 Strain distribution after strengthening – post-tensioned slab

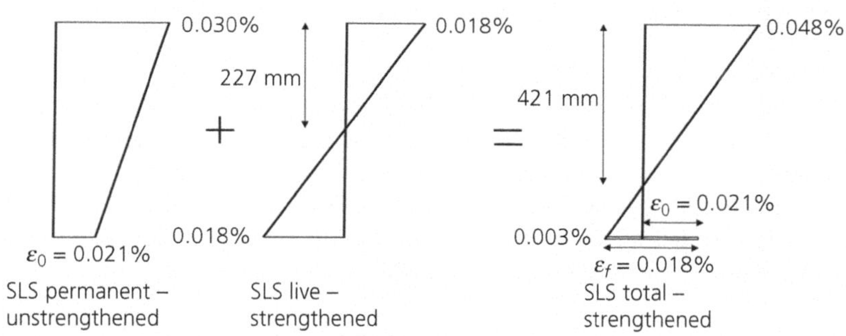

Calculate SLS stresses in strengthened section

The solution can be found by solving the two equilibrium equations in order to find the neutral axis and the strain in the concrete top surface under the SLS live load.

$$\begin{cases} \sum N = 0 \rightarrow F_c - F_{ct} - F_f = 0 \\ \sum M = 0 \rightarrow F_c \dfrac{2x}{3} - F_{ct} \dfrac{2(450-x)}{3} - F_f \left(df - x\right) = 450 \end{cases}$$

The force equilibrium equation can be rewritten as

$$x = \frac{2 \times 2000 \times 450 \times E_{cm}\varepsilon_c + 2A_f E_f \varepsilon_c}{-2A_f E_f \varepsilon_c d_f - 2000 \times 450^2 \times E_{cm}\varepsilon_c}$$

By substituting this in the moment equilibrium equation (or using a simple goal seek algorithm), the strain distribution under the SLS live bending moment in Figure 7.6 is obtained. The total strains are calculated by combining the SLS live strains with the strains at strengthening. The resulting strain in the CFRP is relatively low, demonstrating that in this case the design is driven by stiffness.

The corresponding total stresses in concrete at SLS are as follows.

Top surface

$$\sigma_c = 0.00030 \times E_{c,eff} + 0.00018 \times E_{cm} = 12.06 \text{ MPa} < 12.5 \text{ MPa} - \text{check is ok.}$$

Bottom surface

$$\sigma_{ct} = 0.00021 \times E_{c,eff} + 0.00018 \times E_{cm} = 2.64 \text{ MPa} < 2.65 \text{ MPa} - \text{check is ok.}$$

Crack width

Annex J in *BS EN 1992-1-1:2023* is silent in relation to methods that may be used for determination of crack widths. TR55 (Concrete Society, 2013) indicates that CFRP may be taken into account by transforming the material into either an equivalent reinforcement or concrete area. In undertaking this approach, pre- and post-strengthened crack widths need to be determined and superimposed.

The TR55 approach inherently assumes that there is no bond slip between the concrete and the CFRP. In reality, this is not true as there are different rates of slip between concrete and reinforcement and between concrete and CFRP, and a new crack can form with a different spacing after the CFRP is applied.

ISO14484:2013 (ISO, 2013) specifies that the bond characteristics between CFRP and concrete need to be considered – that is, the slip at the interface needs to be considered. fib Bulletin No. 90 (fib, 2019) provides methods which can account for slip at the interface. It is important to note that the fib Bulletin No. 90 approach was calibrated against BS EN 1992-1-1:2004 (BSI, 2004a), meaning that if the fib Bulletin No. 90 method is applied, it should be used with the provisions in BS EN1992-1-1:2004 to ensure the correct crack width value is obtained.

Crack widths would not normally govern the design of CFRP strengthened sections. In addition, the application of adhesives and the strengthening material itself can reduce both durability concerns and the visibility of cracks to users of structures, which arguably may justify a relaxation of maximum crack width limits set in Section 9 of *BS EN 1992-1-1:2023.*

Given the considerations above, the examples that follow in this chapter are based on the refined method in *BS EN 1992-1-1:2023* for determining crack widths and adopt the TR55 approach of transforming the CFRP into steel. Should refined treatment of crack widths be needed in design, the fib Bulletin No. 90 methods are probably the most reliable alternative that may be used. An example of the use of the fib Bulletin No. 90 method is provided in Annex C.

It is also worth mentioning that the method for calculation of $A_{c,eff}$ used in this book assumes that the placement of CFRP strengthening will be such that it does not impact the effectiveness of the zone of tension stiffening in a reinforced or prestressed concrete member. In routine examples, such as beams in flexure strengthened on the soffit, this assumption should generally hold.

Example EBR Serviceability 3: Crack width

Check crack widths for the section in Example EBR Serviceability 1, assuming that the maximum allowable crack width is 0.4 mm.

The equivalent section will again be used for crack width checks.

From *BS EN 1992-1-1:2023*, Formula 9.8

$$w_{k,cal} \le k_w k_{\frac{1}{r}} s_{r,m,cal} \left(\varepsilon_{sm} - \varepsilon_{cm}\right)$$

The slab is in bending, therefore Formula 9.9 applies for determining $k_{\frac{1}{r}}$

$$k_{\frac{1}{r}} = \frac{h - x}{h - a_{y,i} - x}$$

With reference to figure 9.1 in the standard, $a_{y,i} = 58$ mm (taken as the distance from soffit to bar centreline)

$h = 250$ mm

$x = 62.2$ mm post strengthening

Which gives

$k_{\frac{1}{r}} = 1.45$

$k_w = 1.7$ per *BS EN 1992-1-1:2023*, 9.2.3(2)

As per *BS EN 1992-1-1:2023*, Formula 9.11

$$\varepsilon_{sm} - \varepsilon_{cm} = \frac{\sigma_s - k_t \dfrac{f_{ct,eff}}{\rho_{p,eff}}\left(1 + \alpha_e \rho_{p,eff}\right)}{E_s} \ge \left(1 - k_t\right)\frac{\sigma_s}{E_s}$$

As there is no prestressing, $\rho_{p,eff}$ is defined in Formula 9.12 as

$$\rho_{p,eff} = \frac{A_{equiv}}{A_{c,eff}}$$

A_{equiv} is used in the above equation in place of the steel area to account for the presence of the CFRP.

To determine $A_{c,eff}$, *BS EN 1992-1-1:2023*, Figure 9.3 (a) and (b) is used. Note that for simplicity, for this calculation ϕ is taken as the reinforcement bar diameter (16 mm).

Therefore:

$$h_{c,eff} = min\left(a_y + 5\phi; 10\phi; 3.5a_y; h - x; \frac{h}{2}\right)$$

$$h_{c,eff} = min\left(138; 160; 203; 188; 125\right) = 125 \text{ mm}$$

$b = 1000$ mm

$$A_{c,eff} = 125000 \text{ mm}^2$$

$$\rho_{p,eff} = \frac{2336}{125000} = 0.019$$

$k_t = 0.4$ for sustained loads or many cycles of repeated loading

As stated in *BS EN 1992-1-1:2023*, 9.1.3: $f_{ct,eff} = f_{ctm} = 3.2$ MPa

σ_{equiv} is to be used in place of σ_s, giving

$$\sigma_{equiv} = \sigma_{s,pre} + \varepsilon_{equiv} E_s = \sigma_{s,pre} + \frac{\sigma_{c,post}}{E_c}\left(\frac{d_{equiv} - x}{x}\right) E_s$$

$$= 150.9 + \frac{12.3}{33282} \times \left(\frac{200.1 - 62.2}{62.2}\right) \times 200000 = 314.8 \text{ MPa}$$

This value of stress has been calculated noting that the total value of force in the tension steel and CFRP will be equal to the concrete compressive force because there is no applied axial force on the section.

$$\varepsilon_{sm} - \varepsilon_{cm} = \frac{\sigma_{equiv} - k_t \dfrac{f_{ct,eff}}{\rho_{p,eff}}\left(1 + \alpha_e \rho_{p,eff}\right)}{E_s}$$

$$= \frac{314.8 - 0.4 \times \dfrac{3.2}{0.019}\left(1 + \dfrac{200000}{13441} \times 0.019\right)}{200000} = 0.001143$$

$$\left(1 - k_t\right)\frac{\sigma_{equiv}}{E_s} = \left(1 - 0.4\right) \times \frac{314.8}{200000} = 0.00094 < 0.001143$$

Therefore

$$\varepsilon_{sm} - \varepsilon_{cm} = 1.143 \times 10^{-3}$$

$$s_{rm,cal} = 1.5c + \frac{k_{fl} k_b}{7.2} \frac{\phi}{\rho_{p,eff}} \le \frac{1.3}{k_w}\left(h - x\right)$$

c is the clear cover to the bar, which is $c = 50$ mm

k_{fl} may be calculated using Formula 9.16 as the section is in pure bending: $k_{fl} = \dfrac{h - h_{c,eff}}{h}$

$$k_{fl} = \frac{250 - 125}{250} = 0.5$$

$k_b = 0.9$ for good bond conditions

$$s_{rm,cal} = 1.5 \times 50 + \frac{0.5 \times 0.9}{7.2} \frac{16}{0.019} = 128.5 \text{ mm} \leq \frac{1.3}{k_w}(h - x) = \frac{1.3}{1.7} \times (250 - 62.2) = 143.6 \text{ mm}$$

Therefore

$$w_{k,cal} = k_w k_{\frac{1}{r}} s_{r,m,cal} (\varepsilon_{sm} - \varepsilon_{cm}) = 1.7 \times 1.45 \times 128.5 \times 1.143 \times 10^{-3} = 0.36 \text{ mm} < 0.4 \text{ mm}$$

The crack width here is potentially conservative. One could determine the pre and post strengthening crack widths and superimpose. In this case, assuming that post-strengthening loads were all transient, E_{cm} could be used in place of $E_{c,eff}$ to determine $\varepsilon_{sm} - \varepsilon_{cm}$ and k_t could be taken as 0.6 rather than 0.4, which would yield a lower value of the crack width. Of course, it is still possible that slip between the CFRP and the concrete substrate could result in larger crack widths. Should it be necessary to obtain an accurate estimate including this effect, Annex C provides an example application of the fib90 (2019) method, which could be used.

Deflection control

ULS strengthening may only require relatively low amounts of CFRP relative to the reinforcement or prestressing steel to be used in order to achieve the required member resistance. If there is a deflection issue in the structure, this may necessitate that additional CFRP be provided for deflection control in addition to what is required for ULS strengthening.

BS EN 1992-1-1:2023, J.9(1) clarifies that the modular ratio approach, already described in this chapter for stress limits, may be used for determining deflections. This would mean transforming the CFRP into an equivalent steel area and using the provisions of *BS EN 1992-1-1:2023*.

fib Bulletin No. 90 (fib, 2019) outlines that using the general method for deflection calculation can produce reliable results for EBR CFRP but highlights that a significant experimental validation of the approach has not been undertaken for NSM CFRP systems. There are two potential issues with NSM systems cited in the fib Bulletin No. 90 guidance document.

(a) The adhesive thickness can be relatively high in comparison to EBR systems, meaning that there is a higher slip at the interface. This brings into question the basic assumption of no slip at the interface between CFRP and concrete in transforming CFRP to another material in the structural system.

(b) Ensuring that adhesive is applied uniformly throughout the full length of a slot cut into the concrete structure, into which to adhere the NSM system, is difficult. This can lead to even larger values of slip at the interface.

If using NSM systems in an application where deflection may govern, consultation with the supplier to overcome the known issues should be undertaken to avoid any ongoing deflection issues in the strengthened structure.

Example EBR serviceability 4: Deflection

Find the deflection at midspan for the EBR CFRP strengthened beam illustrated in Figure 7.7.

Assume the properties in Table 7.2, ignoring steel in compression.

Figure 7.7 Strengthened beam for calculating deflection

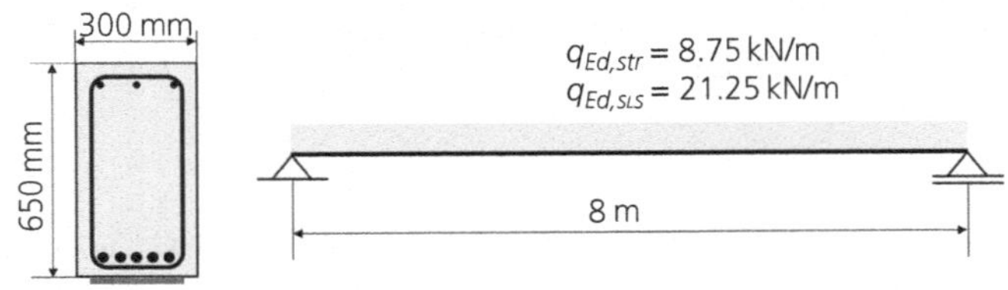

Table 7.2 Material properties for Example EBR Serviceability: Deflection

Reinforcing Steel Properties		CFRP Properties		Concrete Properties	
d_{st}	576 mm	b_f	250 mm	f_{cm}	38 MPa
ϕ	24 mm	Layers	2	f_{ck}	30 MPa
Number	5	A_f	64.5 mm²	f_{cd}	17 MPa
A_{st}	2262 mm²	f_{fuk}	3600 MPa	f_{ctm}	2.9 MPa
f_{yk}	400 MPa	f_{fud}	1800 MPa	$f_{ctk,0.05}$	2 MPa
f_{yd}	348 MPa	E_f	230 000 MPa	f_{ctd}	0.9 MPa
E_s	200 000 MPa	ε_{fuk}	1.57 %	E_{cm}	31 939 MPa
ε_{uk}	5% (Ductility class B)			$E_{c,eff}$	12 898 MPa

The mean deflection may be determined from *BS EN 1992-1-1:2023*, Formula 9.28.

$$a_\delta = (1-\zeta)a_I + \zeta a_{II}$$

Where a_I and a_{II} are the deflections calculated for the uncracked and fully cracked conditions, respectively, and ζ is the distribution coefficient.

Maximum moment at SLS: $M_{Ed,SLS} = 170$ kNm

Moment at strengthening: $M_{Ed,str} = 70$ kNm

Additional moment post-strengthening: $M_{post} = 170 - 70 = 100$ kNm

For cracked sections under sustained load or many cycles of repeated loading, refer to *BS EN 1992-1-1:2023*, Formula 9.29.

$$\zeta = 1 - \beta_t \left(\frac{\sigma_{sr}}{\sigma_s}\right)^2$$

σ_s is the highest stress having occurred up to the moment being analysed in the tension reinforcement, calculated on the basis of a cracked section, and

σ_{sr} is the stress in the tension reinforcement calculated on the basis of a cracked section under the loading conditions causing first cracking.

For a simply supported beam under uniformly distributed load (UDL), q_{Ed}, and applied bending moment $M = \frac{ql^2}{8}$, the calculated deflection will be

$$\alpha_\delta = \frac{5q_{Ed}l^4}{384EI} = \frac{40\,Ml^2}{384EI}$$

Consider the deflection as the superposition of two parts.

- The deflection at the time of strengthening, using unstrengthened section properties.
- The deflection at the time of strengthening, using strengthened section properties.

Pre-strengthening

$$a_{\delta,pre} = (1-\xi)a_{I,pre} + \xi a_{II,pre}$$

All pre-strengthening effects are longterm, i.e. $E_{c,eff} = 12898$ MPa is used, and

$\beta_t = 0.5$ for sustained loads

Calculate the neutral axis and moment of inertia at strengthening, assuming cracked section properties.

$$x_{pre} = \frac{-A_{st}E_s + \sqrt{(A_{st}E_s)^2 + 2bA_{st}E_sE_{c,eff}d_{st}}}{bE_{c,eff}}$$

$$= \frac{-2262 \times 200000 + \sqrt{(2262 \times 200000)^2 + 2 \times 300 \times 2262 \times 200000 \times 12898 \times 576}}{1000 \times 12898}$$

$$= 268.3 \text{ mm}$$

$$I_{cr,pre} = \frac{bx_{pre}^3}{3}\frac{E_{c,eff}}{E_s} + A_{st}(d_{st} - x_{pre})^2 = \frac{300 \times 268.3^3}{3} \times \frac{12898}{200000}$$

$$+ 2262 \times (576 - 268.3)^2 = 33.9 \times 10^7 \text{ mm}^4$$

The gross second moment of area using long-term properties for the concrete in steel units is

$$I_g = \frac{bh^3}{12}\frac{E_{c,eff}}{E_s} = \frac{300 \times 650^3}{12}\frac{12898}{200000} = 44.3 \times 10^7 \text{ mm}^4$$

The cracking moment may be calculated using $f_{ct,eff} = f_{ctm,fl}$ as per *BS EN 1992-1-1:2023*, Formula 9.30.

$$f_{ctm,fl} = max\left[\left(1.6 - \frac{h}{1000}\right)f_{ctm}; f_{ctm}\right] = max\left[\left(1.6 - \frac{650}{1000}\right) \times 2.9; 2.9\right] = 2.9 \text{ MPa}$$

$$M_{cr} = f_{ctm,fl}\frac{bh^2}{6} = 2.9 \times \frac{300 \times 650^2}{6} \times 10^{-6} = 61.3\ \text{kNm}$$

The stress in the reinforcement at strengthening based on cracked section properties is

$$\sigma_s = \frac{M_{Ed,str}\left(d_{st} - x_{pre}\right)}{I_{cr,pre}} = \frac{70 \times 10^6 \times \left(576 - 268.3\right)}{33.9 \times 10^7} = 63.6\ \text{MPa}$$

Similarly, the stress at cracking using cracked section properties is

$$\sigma_{sr} = \frac{M_{cr}\left(d_{st} - x_{pre}\right)}{I_{cr,pre}} = \frac{61.3 \times 10^6 \times \left(576 - 268.3\right)}{33.9 \times 10^7} = 55.7\ \text{MPa}$$

$$\zeta = 1 - \beta_t\left(\frac{\sigma_{sr}}{\sigma_s}\right)^2 = 1 - 0.5 \times \left(\frac{55.7}{63.6}\right)^2 = 0.62$$

Noting that the moment of inertia values are in steel units, a_I is calculated as

$$a_{I,pre} = \frac{40M_{Ed,str}l^2}{384E_sI_g} = \frac{40 \times 70 \times 10^6 \times 8000^2}{384 \times 200\,000 \times 44.3 \times 10^7} = 5.3\ \text{mm}$$

Noting again that the moment of inertia values are in steel units, a_{II} is calculated as

$$a_{II,pre} = \frac{40M_{Ed,str}l^2}{384E_sI_{cr,pre}} = \frac{40 \times 70 \times 10^6 \times 8000^2}{384 \times 200\,000 \times 33.9 \times 10^7} = 6.9\ \text{mm}$$

$$a_{\delta,pre} = \left(1 - 0.62\right) \times 5.3 + 0.62 \times 6.9 = 6.3\ \text{mm}$$

Post-strengthening

$$a_{\delta,post} = \left(1 - \xi\right)a_{I,post} + \xi a_{II,post}$$

All post-strengthening effects are shortterm, i.e. $E_{cm} = 31939$ MPa is used, and

$$\beta_t = 1.0$$

$$A_{equiv} = A_{st} + A_f\frac{E_f}{E_s} = 2262 + 64.5 \times \frac{230\,000}{200\,000} = 2336\ \text{mm}^2$$

$$d_{equiv} = \frac{d_fE_fA_f + d_{st}E_sA_{st}}{E_fA_f + E_sA_{st}} = \frac{650 \times 230\,000 \times 64.5 + 576 \times 200\,000 \times 2262}{230\,000 \times 64.5 + 200\,000 \times 2262} = 578.3\ \text{mm}$$

$$x = \frac{-A_{equiv}E_s + \sqrt{\left(A_{equiv}E_s\right)^2 + 2bA_{equiv}E_sE_{cm}\text{d}_{equiv}}}{bE_{cm}}$$

$$x = \frac{-2336 \times 200000 + \sqrt{(2336 \times 200000)^2 + 2 \times 300 \times 2336 \times 200000 \times 31939 \times 578.3}}{300 \times 31939}$$

$$x = 193.7 \text{ mm}$$

$$I_{cr,post} = \frac{bx^3}{3}\frac{E_{cm}}{E_s} + A_{st}(d_{st} - x)^2 + A_f\frac{E_f}{E_s}(d_f - x)^2$$

$$I_{cr,post} = \frac{300 \times 193.7^3}{3} \times \frac{31939}{200000} + 2262 \times (576 - 193.7)^2 + 64.5 \times \frac{230000}{200000} \times (650 - 193.7)^2$$

$$= 46.2 \times 10^7 \text{ mm}^4$$

Using Equation 7.5, noting that the moment of inertia is in steel units, the stress in the reinforcement at 170 kNm is

$$\sigma_s = \sigma_{s,pre} + \sigma_{s,post} = \frac{M_{Ed,str}(d_{st} - x_{pre})}{I_{cr,pre}} + \frac{M_{post}(d_{st} - x)}{I_{cr,post}}$$

$$= \frac{70 \times 10^6 \times (576 - 268.3)}{33.9 \times 10^7} + \frac{100 \times 10^6 \times (576 - 193.7)}{46.2 \times 10^7} = 146.3 \text{ MPa}$$

For the deflection post-strengthening

$$\zeta = 1 - \beta_t\left(\frac{\sigma_{sr}}{\sigma_s}\right)^2 = 1 - 1.0 \times \left(\frac{55.7}{147.1}\right)^2 = 0.86$$

The gross second moment of area in steel units is:

$$I_g = \frac{bh^3}{12}\frac{E_{cm}}{E_s} = \frac{300 \times 650^3}{12}\frac{31\,939}{200\,000} = 109.6 \times 10^7 \text{ mm}^4$$

Noting that the moment of inertia values are in steel units, a_I is calculated as

$$a_I = \frac{40M_{post}l^2}{384E_sI_g} = \frac{40 \times 100 \times 10^6 \times 8000^2}{384 \times 200000 \times 109.6 \times 10^7} = 3 \text{ mm}$$

Noting again that the moment of inertia values are in steel units, a_{II} is calculated as

$$a_{II} = \frac{40M_{post}l^2}{384E_sI_{cr,post}} = \frac{40 \times 100 \times 10^6 \times 8000^2}{384 \times 200000 \times 46.2 \times 10^7} = 7.2 \text{ mm}$$

$$a_{\delta,post} = (1 - 0.86) \times 3 + 0.86 \times 7.2 = 6.6 \text{ mm}$$

Total deflection

$$a_\delta = a_{\delta,pre} + a_{\delta,post} = 6.3 + 6.6 = 12.9 \text{ mm}$$

Kaloyana Kostova and Craig Giaccio
ISBN 978-1-83549-935-1
https://doi.org/10.1108/978-1-83549-934-420251009

Chapter 8
Fatigue

Introduction

The provisions in *BS EN 1992-1-1:2023* (BSI, 2023b) for fatigue in CFRP are relatively new and detailed, compared with other standards that have been produced in the past. The CFRP material and the adhesive, which bonds it to the concrete surface, have very good fatigue characteristics, hence a check is not included of material or adhesive failure in *BS EN 1992-1-1:2023*. Fatigue in ABR CFRP is governed by degradation of the bond at the interface with the concrete substrate onto which it is bonded under repeated loading.

Owing to the different bond properties for NSM and EBR systems, the provisions are different for each case. The provisions in *BS EN 1992-1-1:2023* are more detailed for EBR than they are for NSM systems, owing to the more substantial databases used to derive provisions.

EBR CFRP systems

The approach in *BS EN 1992-1-1:2023* evaluates the change in bond force, $\Delta F_{fEd,fat}$, between cracks spaced at $s_{cr,min}$, in an EBR CFRP system under a fatigue action combination, against limiting criteria. The approach to determining $\Delta F_{fEd,fat}$ is illustrated in Figure 8.1.

The code provides two methods for determining the fatigue resistance, a simplified and a refined method. The simplified method is first applied to check if the stress range in the CFRP is below a certain limit. If the simplified method check cannot be satisfied, the refined method should be used. The refined method also enables the maximum number of cycles to be determined for a given stress range in the CFRP, as explained in more detail in this section.

Simplified fatigue design method for EBR CFRP systems

A simplified stress-based check is provided in *BS EN 1992-1-1:2023*. It is effectively a *deemed to comply* approach which requires checking the condition stated in Equation 8.1. This basic check enables damage calculations to be omitted if it can be demonstrated that the stress in the bond between the concrete and CFRP does not exceed the initial linear elastic response.

$$\Delta F_{fRd,fat1} \geq max \begin{cases} b_f t_f \Delta f_{fEd,max} \\ F_{fEd,cr} \end{cases} \tag{8.1}$$

Where, as shown in Equation 8.2

$$\Delta F_{fRd,fat1} = 0.35 f_{ctm,surf}^{\frac{1}{4}} f_{bfRd} A_f \tag{8.2}$$

Figure 8.1 CFRP fatigue stress range under a combination of actions

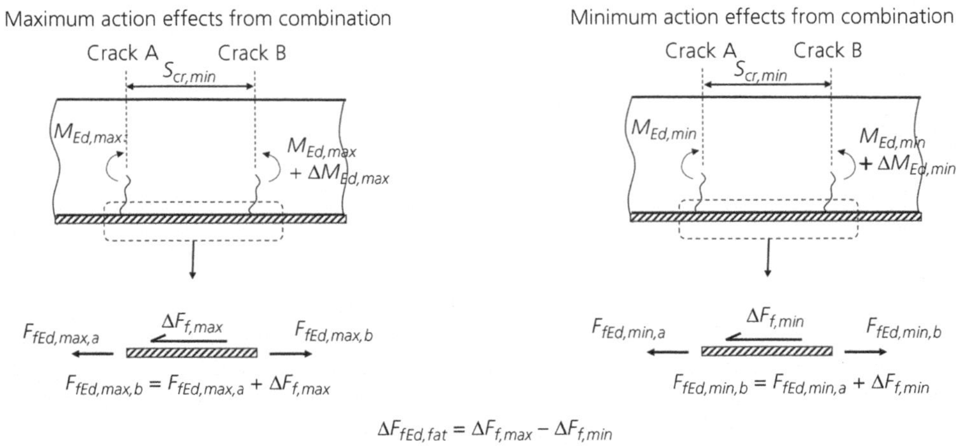

Where

f_{bfRd} is defined in *BS EN 1992-1-1:2023*, J.10.1(1), which is discussed further below.

$$\Delta f_{fEd,max} = \frac{\Delta F_{fEd,max}}{A_f}$$

$\Delta F_{fEd,max}$ is the maximum value of $\Delta F_{fEd,fat}$ between two cracks, shown in Figure 8.1, under the fatigue load combination being considered.

$F_{fEd,cr}$ is the stress in the CFRP at the first crack in the strengthened area of the member.

To determine the limiting bond resistance, f_{bfRd}, for the simplified method, *BS EN 1992-1-1:2023*, J.01.1(1) states that this is the 'limiting design strength of the bond in the area being considered calculated in accordance with J.11'.

For fatigue, the governing bond resistance is the bond component of the intermediate crack debonding resistance ($\Delta f_{fk,B}$ from Equation 3.12) outside of the anchorage region. As such, there appears to be some inconsistency in terminology in the code between *BS EN 1992-1-1:2023*, J.10.1(1) and 3.5.2. This inconsistency will likely be clarified via Complentary Information documents issued in the UK. The limiting bond strength for use in Equation 8.2, where fatigue checks are undertaken outside of the anchorage region, should be modified to f_{bfRd}^* to avoid confusion with the symbol f_{bfRd} as defined in *BS EN 1992-1-1:2023*, 3.5.2, may be determined using Equation 8.3.

$$f_{bfRd}^* = \frac{\sqrt{\eta_{cc} k_{tc} k_{tt} \Delta f_{fk,B}}}{\gamma_{BA}} \quad (8.3)$$

Where all parameters are defined in Chapter 3 of this book.

Requirements for checking fatigue in the anchorage area are not stated in *BS EN 1992-1-1:2023*, J.10. It is suggested here that fatigue checks extend for the full length of the anchorage. Where fatigue checks are undertaken in the anchorage region, the value of the anchorage resistance (f_{bfRd} defined in Equation 3.8) should also be used in Equation 8.2 to ensure that the governing bond resistance is used in this Equation.

Refined fatigue design method for EBR CFRP systems

Where the condition in Equation 8.2 (*BS EN 1992-1-1:2023*, Formula J.25) cannot be satisfied, a method to determine the maximum change in CFRP stress under a given number of cycles has been included in the standard.

The code formulation determines the maximum change in the bond force resisted between cracks before fatigue failure of the adhesive bond between the CFRP and concrete occurs, for a given number of stress cycles, N^*. The adhesive bond component of the bond resistance is determined in accordance with Equation 3.12 (*BS EN 1992-1-1:2023*, Formula J.48). The basic requirement for the fatigue limit state is shown in Equation 8.4.

$$\Delta F_{fEd,fat} \leq \Delta F_{fRd,fat2} \tag{8.4}$$

Where

$\Delta F_{fEd,fat}$ is the corresponding change in force difference between cracks in the CFRP system subjected to N^* cycles, defined in Figure 8.1.

$$F_{fRd,fat2} = \alpha_{fat2} \frac{\Delta F_{fk,B}}{\gamma_{BA}} \tag{8.5}$$

Where

$\Delta F_{fk,B}$ is the adhesive resistance between cracks determined using Equation 8.6.

$$\Delta F_{fk,B} = \Delta f_{fk,B} A_f \tag{8.6}$$

Where $\Delta f_{fk,B}$ is defined in Equation 3.12 (*BS EN 1992-1-1:2023*, Formula J.48), a_{fat2} is defined in Equation 8.7.

$$\alpha_{fat2} = -c_{fat} \frac{\Delta F_{f,max}}{\Delta F_{fRd}} + c_{fat} \tag{8.7}$$

Where, c_{fat} as shown in Equation 8.8

$$c_{fat} = 0.35 \left(\frac{N^*}{2 \times 10^6} \right)^{-\frac{1}{k_{f3}}} \tag{8.8}$$

N^* is the number of cycles for a given stress (force) cycle.

ΔF_{fRd} is the intermediate crack debonding resistance between cracks spaced at $s_{cr,min}$ calculated using Equation 8.9, where Δf_{fRd} is determined using Equation 3.11 (*BS EN 1992-1-1:2023*, Formula J.47).

$$\Delta F_{fRd} = \Delta f_{fRd} b_f t_f \tag{8.9}$$

$\Delta F_{f,max}$ is the maximum value of the difference in force in the EBR CFRP system between cracks spaced at $s_{cr,min}$ under the load cycle from the fatigue combination being considered.

γ_{BA} is the partial factor for bond strength from *BS EN 1992-1-1:2023*, Table J.1.

$$k_{f3} = \begin{cases} 23.2 \text{ for } N^* < 2 \times 10^6 \\ 45.4 \text{ for } N^* \geq 2 \times 10^6 \end{cases}$$

Combining the equations above, the maximum value of the change in force difference between cracks spaced at $s_{cr,min}$ for a given number of cycles N^* can be evaluated as shown in Equation 8.10.

$$\Delta F_{fRd,fat2} = 0.35 \left(\frac{N^*}{2 \times 10^6} \right)^{-\frac{1}{k_{f3}}} \left(1 - \frac{\Delta F_{f,max}}{\Delta F_{fRd}} \right) \left(\frac{\Delta F_{fk,B}}{\gamma_{BA}} \right) \tag{8.10}$$

To find the maximum number of cycles that can be resisted by the CFRP for a given change in force difference between two cracks, the condition in Equation 8.11 can be applied.

$$\Delta F_{fEd,fat} = \Delta F_{fRd,fat2} \tag{8.11}$$

Therefore, as shown in Equation 8.12

$$\Delta F_{fEd,fat} = 0.35 \left(\frac{N^*}{2 \times 10^6} \right)^{-\frac{1}{k_{f3}}} \left(1 - \frac{\Delta F_{f,max}}{\Delta F_{fRd}} \right) \left(\frac{\Delta F_{fk,B}}{\gamma_{BA}} \right) \tag{8.12}$$

This can be manipulated to produce the maximum number of cycles, N^*_{max}, that the system can sustain under the change in CFRP force difference for the fatigue load combination being considered. This is shown in Equation 8.13.

$$N^*_{max} = 2 \times 10^6 \left[\frac{2.86 \Delta F_{fEd,fat}}{\left(1 - \frac{\Delta F_{f,max}}{\Delta F_{fRd}} \right) \left(\frac{\Delta F_{fk,B}}{\gamma_{BA}} \right)} \right]^{-k_{f3}} \tag{8.13}$$

To ensure fatigue failure does not occur in the anchorage region, it would be prudent to continue these fatigue checks into the anchorage region.

Determination of Force difference

As stated in the section on intermediate crack debonding in Chapter 5, the difference in forces between cracks is most accurately calculated using an iterative strain compatibility-based method. If such a method is not available, the simplified method in Chapter 5 may be used. It ensures a smoother transition in the area of yielding reinforcement than other available methods.

Example EBR Fatigue 1: Simplified method for fatigue evaluation

Determine if detailed fatigue checks are required for the beam strengthened with CFRP shown in Figure 8.2. Assume the following:

- Concrete class C30/37
- It is known that the beam was cast such that the strengthening is applied at the bottom surface of the beam with respect to the casting position
- $k_{sys,b1} = k_{sys,b2} = k_{sys,b3} = 1.0$
- $\Delta f_{fEd,max} b_f t_f = 15.2$ kN

The cross-sectional area of the fibres per metre of a single layer of CF sheet given by the supplier is 129 mm^2/m. Therefore, for a metre width of a CF sheet $A_{f,1} = 129$ mm^2 per single layer

- The maximum force in the crack closest to the point of contraflexure, within the area of strengthening under the fatigue combination being considered, has been determined by an iterative cross-sectional analysis to be $F_{fEd,cr} = 20.1$ kN.

Assume the properties given in Table 8.1 and only undertake fatigue checks outside of the anchorage region.

Figure 8.2 Strengthened beam for calculating SN curve

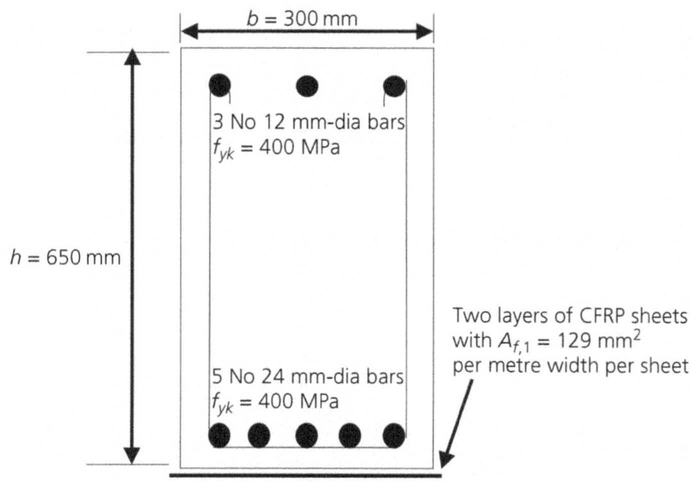

Table 8.1 Material properties – Example EBR Fatigue 1

Reinforcing Steel Properties		CFRP Properties		Concrete Properties	
d_{st}	576 mm	E_f	230 000 MPa	Grade	C30/37
ϕ	24 mm	f_{fuk}	3600 MPa	f_{cm}	38 MPa
Number	5			f_{ck}	30 MPa
A_{st}	2261 mm^2			f_{cd}	17 MPa
f_{yk}	400 MPa			f_{ctm}	2.9 MPa
f_{yd}	348 MPa			$f_{ctk,0.05}$	2.0 MPa
E_s	200 000 MPa			f_{ctd}	0.9 MPa
ε_{uk}	5% (Ductility class B)			E_{cm}	31 939 MPa

Reinforcing bars are ribbed and in good bond conditions

To confirm if detailed fatigue checks are required, the condition in Equation 8.1 and Equation 8.2 need to be met as follows.

$$0.35 f_{ctm,surf}{}^{1/4} f_{bfRd} b_f t_f \geq max\begin{cases} b_f t_f \Delta f_{fEd,max} \\ F_{fEd,cr} \end{cases}$$

As indicated in Figure 8.2, the strengthening is full width, therefore $b_f = 300$ mm

Firstly, determine the effective thickness of the system.

$$t_f = n_f{}^{k_f} \frac{A_{f,1}}{b_f} = 2^1 \frac{129}{1000} = 0.258 \text{ mm}$$

Now that A_f is known, $f_{ctm,surf}$, $s_{cr,min}$, and f_{bfRd} need to be determined to assess both conditions of bond fatigue stated in Equation 8.3.

BS EN 1992-1-1:2023, Formula J.34 gives the following expression for $f_{ctm,surf}$

$$f_{ctm,surf} = k_{c,surf} f_{ctm}$$

As the beam is categorised as having 'concrete position: bottom',

$$k_{c,surf} = 0.6 + 0.3\left(\frac{f_{ck}}{60} - 0.2\right)$$

$f_{ck} = 30$ MPa

Therefore

$$k_{c,surf} = 0.6 + 0.3 \times \left(\frac{30}{60} - 0.2\right) = 0.69$$

Therefore

$$f_{ctm,surf} = k_{c,surf} f_{ctm} = 0.69 \times 2.9 = 2.0 \text{ MPa}$$

To determine the relevant bond resistance, the minimum crack spacing is required as shown in Equation 3.9.

$$s_{cr,min} = 1.5 \frac{M_{cr}}{0.85\ h\ F_{bsm}}$$

The cracking moment for a rectangular section can be calculated as follows.

$$M_{cr} = f_{ctm} \frac{I}{h_{max}} = \frac{f_{ctm} I}{\frac{h}{2}}$$

Where

I is the second moment of area of concrete (of the gross section)

$$= \frac{bh^3}{12}$$

h is the overall member depth as shown in Figure 8.2

b is the overall member width as shown in Figure 8.2

Which gives

$$I = \frac{300 \times 650^3}{12} = 6.866 \times 10^9 \text{ mm}^4$$

Therefore

$$M_{cr} = \frac{2.9 \times 6.866 \times 10^9}{\frac{650}{2}} = 61.3 \text{ kNm}$$

The bond stress around each ribbed bar required for determining $s_{cr,min}$ may be calculated using *BS EN 1992-1-1:2023*, Formula J.46 as follows.

$$f_{fbsm} = \kappa_{vb1} 0.43 f_{cm}^{\frac{2}{3}}$$

As the bars are ribbed, κ_{vb1} applies and, given they are cast in good conditions,

$$\kappa_{vb1} = 1.0$$

Therefore

$$f_{fbsm} = 0.43 f_{cm}^{\frac{2}{3}} = 0.43 \times 38^{\frac{2}{3}} = 4.86 \text{ MPa}$$

BS EN 1992-1-1:2023, Formula J.45 is used to determine the bond force per unit length of bars as follows.

$$F_{fbsm} = \sum_{i=1}^{m} n_{s,i} \varnothing_i \pi f_{fbsm}$$

Where

i the number of different bar diameters in the tension zone of a member in flexure

$\varnothing_i$ is the diameter of bar i

$n_{s,i}$ is the number of bars of given diameter $\varnothing_i$

Therefore

$$F_{fbsm} = \sum_{i=1}^{1} n_{s,i} \varnothing_i \pi f_{fbsm} = 5 \times 24 \times \pi \times 4.86 = 1832 \text{ N / mm}$$

Which gives

$$s_{cr,min} = 1.5 \times \frac{61.3 \times 10^6}{0.85 \times 650 \times 1832} = 91 \text{ mm}$$

From *BS EN 1992-1-1:2023*, Formula 5.4

$$\eta_{cc} = \left(\frac{f_{ck,ref}}{f_{ck}}\right)^{\frac{1}{3}} \leq 1.0$$

Which gives $\eta_{cc} = 1.0$

Using the recommended values of k_{tc} and k_{tt} from *BS EN 1992-1-1:2023*, 5.1.6 gives

$$k_{tc} = 0.85$$

$$k_{tt} = 0.7$$

For bond resistance, γ_{BA} is taken from *BS EN 1992-1-1:2023*, Table J.1 for the fatigue design case, therefore

$$\gamma_{BA} = 1.5$$

Determine the bond resistance component of the intermediate crack resistance, $f_{bfRd}{}^*$, as per Equation 8.3.

$$f_{bfRd}{}^* = \frac{\sqrt{\eta_{cc} k_{tc} k_{tt} \Delta f_{fk,B}}}{\gamma_{BA}}$$

$$f_{bfRd}^{*} = 0.84 \times \frac{\sqrt{1 \times 0.85 \times 0.7}}{1.5} \times 1.0 \times \sqrt{38 \times 2} \times \frac{\sqrt{91}}{0.258} = 139.2 \text{ MPa}$$

Therefore, the basic fatigue resistance $\Delta F_{Rd,fat1}$ may be determined from Equation 8.2 using f_{bfRd}^{*}

$$\Delta F_{fRd,fat1} = 0.35 f_{ctm,surf}^{\frac{1}{4}} f_{bfRd}^{*} b_f t_f = 0.35 \times 2^{\frac{1}{4}} \times 139.2 \times 300 \times 0.258 = 4.5 \text{ kN}$$

As given in the example text

$$\Delta F_{fE,equ} = max \begin{cases} \Delta f_{fEd,max} A_f = 15.2 \text{ kN} \\ F_{fEd,cr} = 20.1 \text{ kN} \end{cases} > \Delta F_{fRd,fat1} = 4.5\text{kN}$$

Therefore, a detailed fatigue assessment is required.

Example EBR Fatigue 2: Refined method for single stress range

For the slab in Example Flexural Design 1, calculate the permissible number of stress cycles, assuming the slab has a span of 8 m, is simply supported, and carries a uniformly distributed load (UDL) at strengthening of $q_{Ed,str} = 6.25$ kN/m, and resists a stress range in service generated by a UDL of $q_{Ed,min} = 12.5$ kN/m and $q_{Ed,max} = 20.5$ kN/m.

Using values calculated in Example Flexural Design 1, the following is obtained for intermediate crack debonding.

$$\Delta f_{fk,B} = 0.84 k_{sys,b1} \sqrt{f_{cm} f_{ctm,surf}} \frac{\sqrt{s_{cr,min}}}{t_f} = 0.84 \times 1 \times \sqrt{43 \times 2.3} \times \frac{\sqrt{89}}{1.2} = 65.4 \text{ MPa}$$

$$\Delta_{fk,F} = f_{cm}^{-0.9} \frac{s_{cr,min}^{\frac{4}{3}}}{t_f} = 43^{-0.9} \times \frac{89^{4/3}}{1.2} = 11.2 \text{ MPa}$$

$$\Delta f_{fk,C} = \frac{\kappa_h}{\max(100\,\text{mm}, h)} \cdot \frac{s_{cr,min}^{\frac{1}{3}}}{t_f} = \frac{2000}{250} \cdot \frac{89^{\frac{1}{3}}}{1.2} = 29.7 \text{ MPa}$$

$$\Delta f_{fRd} = \frac{1}{\gamma_{BA}} \times \left[\left(\eta_{cc} k_{tc} k_{tt} \right)^{0.5} \Delta f_{fk,B} + \Delta f_{fk,F} + \Delta f_{fk,C} \right]$$
$$= \frac{1}{1.5} \times \left[\left(1.0 \times 0.85 \times 0.7 \right)^{0.5} \times 65.5 + 11.2 + 29.7 \right] = 60.9 \text{ MPa}$$

$$\Delta F_{fRd} = \Delta f_{fRd} A_f = 60.9 \times 1.2 \times \frac{450}{1000} = 32.9 \text{ kN}$$

$$\Delta F_{fk,B} = \Delta f_{fk,B} A_f = 65.5 \times 1.2 \times \frac{450}{1000} = 35.4 \text{ kN}$$

From Table 8.2, the maximum change in force difference between cracks is

$$\Delta F_{fEd,fat} = \Delta F_{f,max} - \Delta F_{f,min} = 6.46 - 2.83 = 3.63 \text{ kN}$$

Note that the force difference has been calculated where the moment exceeds the cracking moment

From Equation 8.13

$$N^*_{max} = 2\times10^6 \left[\frac{2.86\ \Delta F_{fEd,fat}}{\left(1-\dfrac{\Delta F_{f,max}}{\Delta F_{fRd}}\right)\left(\dfrac{\Delta F_{fk,B}}{\gamma_{BA}}\right)} \right]^{-k_{f3}}$$

Assuming that there are more than 2×10^6 cycles (yielding $k_{f3} = 45.4$) and substituting in the relevant values, the maximum number of cycles will be

$$N^*_{max} = 2\times10^6 \left[\frac{2.86\times3.63}{\left(1-\dfrac{6.46}{32.9}\right)\left(\dfrac{35.4}{1.5}\right)} \right]^{-45.4} = 1.71\times10^{18} \text{ cycles}$$

Example EBR Fatigue 3: Refined method SN curve

Produce an SN curve for the beam described in Example EBR Fatigue 1, assuming that no positive sagging has been observed on site, which could realistically be relied upon for adding resistance to intermediate crack debonding resistance.

To produce an SN curve for the beam in Example EBR Fatigue 1, Equation 8.13 is used to determine the maximum number of stress cycles, N^*_{max}, for a given change in force difference between two cracks, $\Delta F_{fEd,fat}$. This is as follows.

$$N^*_{max} = 2\times10^6 \left[\frac{2.86\Delta F_{fEd,fat}}{\left(1-\dfrac{\Delta F_{f,max}}{\Delta F_{fRd}}\right)\left(\dfrac{\Delta F_{fk,B}}{\gamma_{BA}}\right)} \right]^{-k_{f3}}$$

Using the calculations completed in Example EBR Fatigue 1, the provisions of *BS EN 1992-1-1:2023*, J.11 will be applied to determine the intermediate crack debonding resistance.

As per Equation 8.13, determination of the number of cycles to failure under a given load cycle, the full intermediate crack debonding force resistance, ΔF_{fRd}, is also required, as follows.

$$\Delta F_{fRd} = \Delta f_{fRd} b_f t_f$$

Where Δf_{fRd} is given in Equation 3.11 (*BS EN 1992-1-1:2023*, Formula J.47).

Table 8.2 Tabulated data – Example EBR Fatigue 2 (continued on next page)

	Pre-Str		Min Force						Max Force						
	$M_{Ed,str}$	$\sigma_{s,pre}$	M_{Ed}	σ_s	F_{fEd}	f_{fEd}	ΔF_{fEd}	Δf_{fEd}	M_{Ed}	σ_s	F_{fEd}	f_{fEd}	ΔF_{fEd}	Δf_{fEd}	ΔF_f
Distance	(kNm)	(MPa)	(kNm)	(MPa)	(kN)	(MPa)	(kN)	(MPa)	(kNm)	(MPa)	(kN)	(MPa)	(kN)	(MPa)	
0	0.00		0.00						0.00						
84	2.08	5.98	4.16	10.89	2.94	5.44	2.94	5.44	6.82	17.18	6.69	12.40	6.69	12.40	3.76
173	4.23	12.17	8.46	22.17	5.98	11.07	3.04	5.63	13.88	34.98	13.63	25.24	6.94	12.85	3.89
262	6.34	18.22	12.67	33.20	8.95	16.58	2.97	5.50	20.78	52.37	20.41	37.79	6.78	12.55	3.80
351	8.39	24.13	16.78	43.97	11.85	21.95	2.90	5.38	27.52	69.35	27.03	50.05	6.62	12.26	3.72
440	10.40	29.90	20.79	54.47	14.69	27.20	2.83	5.25	34.10	85.92	33.48	62.01	6.46	11.96	3.63
529	12.35	35.52	24.70	64.72	17.45	32.31	2.76	5.12	40.51	102.09	39.78	73.67	6.30	11.66	3.54
618	14.26	41.01	28.51	74.71	20.14	37.30	2.69	4.99	46.76	117.84	45.92	85.04	6.14	11.37	3.45
707	16.11	46.35	32.23	84.43	22.76	42.16	2.62	4.86	52.85	133.19	51.90	96.12	5.98	11.07	3.36
796	17.92	51.54	35.84	93.90	25.32	46.88	2.55	4.73	58.78	148.13	57.72	106.90	5.82	10.78	3.27
885	19.68	56.60	39.35	103.11	27.80	51.48	2.48	4.60	64.54	162.65	63.38	117.38	5.66	10.48	3.18
974	21.39	61.51	42.77	112.06	30.21	55.95	2.41	4.47	70.14	176.77	68.89	127.57	5.50	10.19	3.09
1063	23.04	66.28	46.09	120.75	32.56	60.29	2.34	4.34	75.58	190.48	74.23	137.46	5.34	9.89	3.00
1152	24.65	70.91	49.31	129.19	34.83	64.50	2.27	4.21	80.86	203.78	79.41	147.06	5.18	9.60	2.91
1241	26.21	75.39	52.42	137.36	37.03	68.58	2.20	4.08	85.98	216.67	84.43	156.36	5.02	9.30	2.82
1330	27.72	79.74	55.44	145.27	39.17	72.53	2.13	3.95	90.93	229.15	89.30	165.37	4.86	9.01	2.73
1419	29.18	83.94	58.37	152.92	41.23	76.35	2.06	3.82	95.72	241.22	94.00	174.08	4.70	8.71	2.64
1508	30.59	88.00	61.19	160.32	43.22	80.04	1.99	3.69	100.35	252.89	98.55	182.50	4.54	8.42	2.55
1597	31.95	91.91	63.91	167.45	45.15	83.60	1.92	3.56	104.81	264.14	102.93	190.62	4.39	8.12	2.46
1686	33.27	95.68	66.53	174.32	47.00	87.04	1.85	3.43	109.12	274.98	107.16	198.44	4.23	7.83	2.37
1775	34.53	99.32	69.06	180.94	48.78	90.34	1.78	3.30	113.26	285.42	111.23	205.97	4.07	7.53	2.28
1864	35.74	102.80	71.48	187.30	50.50	93.51	1.71	3.17	117.23	295.44	115.13	213.21	3.91	7.24	2.19
1953	36.91	106.15	73.81	193.39	52.14	96.56	1.64	3.04	121.05	305.06	118.88	220.15	3.75	6.94	2.10

Table 8.2 Continued

	Pre-Str		Min Force						Max Force						
Distance	$M_{Ed,str}$ (kNm)	$\sigma_{s,pre}$ (MPa)	M_{Ed} (kNm)	σ_s (MPa)	F_{fEd} (kN)	f_{fEd} (MPa)	ΔF_{fEd} (kN)	Δf_{fEd} (MPa)	M_{Ed} (kNm)	σ_s (MPa)	F_{fEd} (kN)	f_{fEd} (MPa)	ΔF_{fEd} (kN)	Δf_{fEd} (MPa)	ΔF_f
2042	38.02	109.35	76.04	199.23	53.71	99.47	1.57	2.91	124.70	314.27	122.47	226.79	3.59	6.64	2.01
2131	39.08	112.42	78.17	204.81	55.22	102.25	1.50	2.78	128.20	323.07	125.90	233.14	3.43	6.35	1.92
2220	40.10	115.34	80.20	210.12	56.65	104.91	1.43	2.66	131.52	331.46	129.17	239.20	3.27	6.05	1.84
2309	41.06	118.11	82.13	215.18	58.02	107.44	1.36	2.53	134.69	339.44	132.28	244.95	3.11	5.76	1.75
2398	41.98	120.75	83.96	219.98	59.31	109.83	1.29	2.40	137.69	347.01	135.23	250.42	2.95	5.46	1.66
2487	42.85	123.24	85.69	224.52	60.53	112.10	1.22	2.27	140.54	354.17	138.02	255.59	2.79	5.17	1.57
2576	43.66	125.59	87.33	228.80	61.69	114.24	1.15	2.14	143.22	360.92	140.65	260.46	2.63	4.87	1.48
2665	44.43	127.79	88.86	232.82	62.77	116.24	1.08	2.01	145.73	367.26	143.12	265.04	2.47	4.58	1.39
2754	45.15	129.86	90.30	236.59	63.79	118.12	1.01	1.88	148.09	373.20	145.43	269.32	2.31	4.28	1.30
2843	45.82	131.78	91.63	240.09	64.73	119.87	0.94	1.75	150.28	378.72	147.58	273.30	2.15	3.99	1.21
2932	46.44	133.56	92.87	243.33	65.60	121.49	0.87	1.62	152.31	383.84	149.58	277.00	1.99	3.69	1.12
3021	47.00	135.20	94.01	246.31	66.41	122.98	0.80	1.49	154.18	388.54	151.41	280.39	1.83	3.40	1.03
3110	47.52	136.69	95.05	249.04	67.14	124.34	0.73	1.36	155.88	392.84	153.09	283.49	1.67	3.10	0.94
3199	47.99	138.05	95.99	251.50	67.81	125.57	0.66	1.23	157.42	396.73	154.60	286.30	1.51	2.81	0.85
3288	48.42	139.26	96.83	253.71	68.40	126.67	0.59	1.10	158.80	400.20	155.96	288.81	1.36	2.51	0.76
3377	48.79	140.33	97.57	255.65	68.93	127.64	0.52	0.97	160.02	403.27	157.15	291.02	1.20	2.21	0.67
3466	49.11	141.25	98.22	257.34	69.38	128.48	0.45	0.84	161.08	405.93	158.19	292.94	1.04	1.92	0.58
3555	49.38	142.03	98.76	258.77	69.77	129.20	0.38	0.71	161.97	408.18	159.07	294.57	0.88	1.62	0.49
3644	49.60	142.68	99.21	259.93	70.08	129.78	0.31	0.58	162.70	410.03	159.78	295.90	0.72	1.33	0.40
3733	49.78	143.17	99.55	260.84	70.33	130.23	0.24	0.45	163.27	411.46	160.34	296.93	0.56	1.03	0.31
3822	49.90	143.53	99.80	261.49	70.50	130.56	0.17	0.32	163.68	412.48	160.74	297.67	0.40	0.74	0.22
3911	49.98	143.74	99.95	261.88	70.61	130.75	0.10	0.19	163.92	413.09	160.98	298.11	0.24	0.44	0.13
4000	50.00	143.81	100.00	262.01	70.64	130.81	0.03	0.06	164.00	413.30	161.06	298.26	0.08	0.15	0.04

As noted in Equation 3.12

$$\Delta f_{fk,B} = 0.84 k_{sys,b1} \sqrt{f_{cm} f_{ctm,surf}} \frac{\sqrt{s_{cr,min}}}{t_f}$$

Which yields

$$\Delta f_{fk,B} = 0.84 \times 1.0 \times \sqrt{38 \times 2} \times \frac{\sqrt{91}}{0.258} = 270.8 \text{ MPa}$$

The term $\Delta F_{fk,B}$ is required for Equation 8.13, and can be determined as follows.

$$\Delta F_{fk,B} = \Delta f_{fk,B} b_f t_f = 270.8 \times 300 \times 0.258 = 21.0 \text{ kN}$$

To determine Δf_{fRd} note that, as the example text stipulates, it can be assumed that positive curvature is not available to generate clamping, therefore the term $\Delta f_{fk,C}$ is set to

$$\Delta f_{fk,C} = 0$$

The friction component of the intermediate crack debonding resistance, $\Delta f_{fk,F}$, is calculated using Equation 3.13 (*BS EN 1992-1-1:2023*, Formula J.49) as follows.

$$\Delta f_{fk,F} = 38^{-0.9} \times \frac{91^{\frac{4}{3}}}{0.258} = 60.1 \text{ MPa}$$

As per Equation 3.11 (*BS EN 1992-1-1:2023*, Formula J.47)

$$\Delta f_{fRd} = \frac{\sqrt{\eta_{cc} \cdot k_{tc} \cdot k_{tt}} \cdot \Delta f_{fk,B} + \Delta f_{fk,F} + \Delta f_{fk,C}}{\gamma_{BA}}$$

Which yields

$$\Delta f_{fRd} = \frac{\sqrt{1 \times 0.85 \times 0.7} \times 270.8 + 60.1 + 0}{1.5} = 179.3 \text{ MPa}$$

The term ΔF_{fRd} is required for Equation 8.13; this can be determined as follows.

$$\Delta F_{fRd} = \Delta f_{fRd} b_f t_f = 179.3 \times 300 \times 0.258 = 13.9 \text{ kN}$$

Substituting all of this into Equation 8.13 yields

$$N^*_{max} = 2 \times 10^6 \left[\frac{2.86 \times \Delta F_{fEd,fat}}{\left(1 - \frac{\Delta F_{f,max}}{13.9}\right) \times \left(\frac{21.0}{1.5}\right)} \right]^{-k_{f3}}$$

Figure 8.3 SN curve for EBR Fatigue Example 3

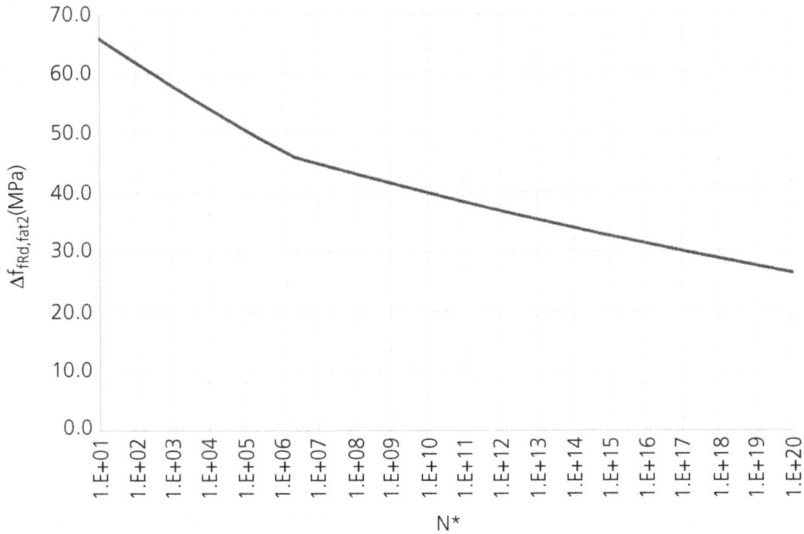

To construct the SN curve, the term $\Delta F_{f,max}$ is required. In this example, it is assumed that the member is cycling such that the force difference between cracks remains constant (akin to a constant amplitude stress cycle), and that the minimum force difference within the stress cycle is $\Delta F_{f,min} = 0.2$ kN.

With this assumption, the SN curve in Figure 8.3 can be produced.

The curve is quite linear when using a logarithmic horizontal axis and shows a distinct change in gradient when the value of k_{f3} changes from 43.4 to 23.2. There is no equivalent condition to a constant amplitude fatigue limit. Noting that the limiting intermediate crack debonding stress is 179 MPa and that the maximum stress for 10 cycles is approximately 65 MPa, it is evident that fatigue could become the governing design in high cycle fatigue applications.

Kaloyana Kostova and Craig Giaccio
ISBN 978-1-83549-935-1
https://doi.org/10.1108/978-1-83549-934-420251008

Chapter 9
Detailing

Introduction
This chapter outlines relevant detailing provisions from *BS EN 1992-1-1:2023* (BSI, 2023b). It also supplies complementary guidance to these provisions where gaps exist in the code.

The first basic requirement is that fibres should be orientated as stated below unless a specific misalignment of the fibres to the line of action of the force being resisted (or confined) has been considered in the design.

- Fibres should align with the principal direction of the tensile forces being resisted for flexure and shear.
- Fibres should be perpendicular to the longitudinal axis for column confinement.

Flexural strengthening with NSM
As shown in Figure 9.1, the NSM CFRP strengthening system requires cutting grooves into the concrete surface. This should only be executed where it can be demonstrated that cutting grooves into the concrete, of sufficient size to achieve minimum dimensions set out in the standard, will not result in existing reinforcement being compromised in any manner. This includes ensuring reinforcement is not cut during installation without a manageable remedial solution. The durability of the member being strengthened should also not be compromised such that the intended design/residual design life cannot be achieved.

In terms of detailing, *BS EN 1992-1-1:2023*, J.12.4 requires that no more than one NSM bar, strip or rod should be used in each slot cut into the structure. The code then provides a series of geometric limitations on spacing that are aimed at ensuring that slots are wide and deep enough to encase the CFRP with sufficient adhesive, as well as ensuring the failure mechanisms considered in the standard can occur. For completeness, the described limitations are reproduced in Figure 9.2 and Table 9.1. In using these provisions, it must be remembered that bond provisions in *BS EN 1992-1-1:2023* only cover NSM CFRP strips. Detailing provisions for square and round bars may be applied where alternative bond resistance provisions are used.

The slot width and depth limits of 1 to 3 mm exist to ensure that there is full bond around the perimeter of the slot. This ensures that enough resin can be applied to achieve good bond without being too thick to cause significant slip between the CFRP and concrete substrate.

As mentioned earlier, spacing limits are imposed to avoid unintended concrete separation, either between bars or at the edge of a concrete member, as follows.

- The centre-to-centre spacing limit ensures that the width of the unreinforced concrete element between slots is at least equal to the depth of each slot. This is to remove the tendency

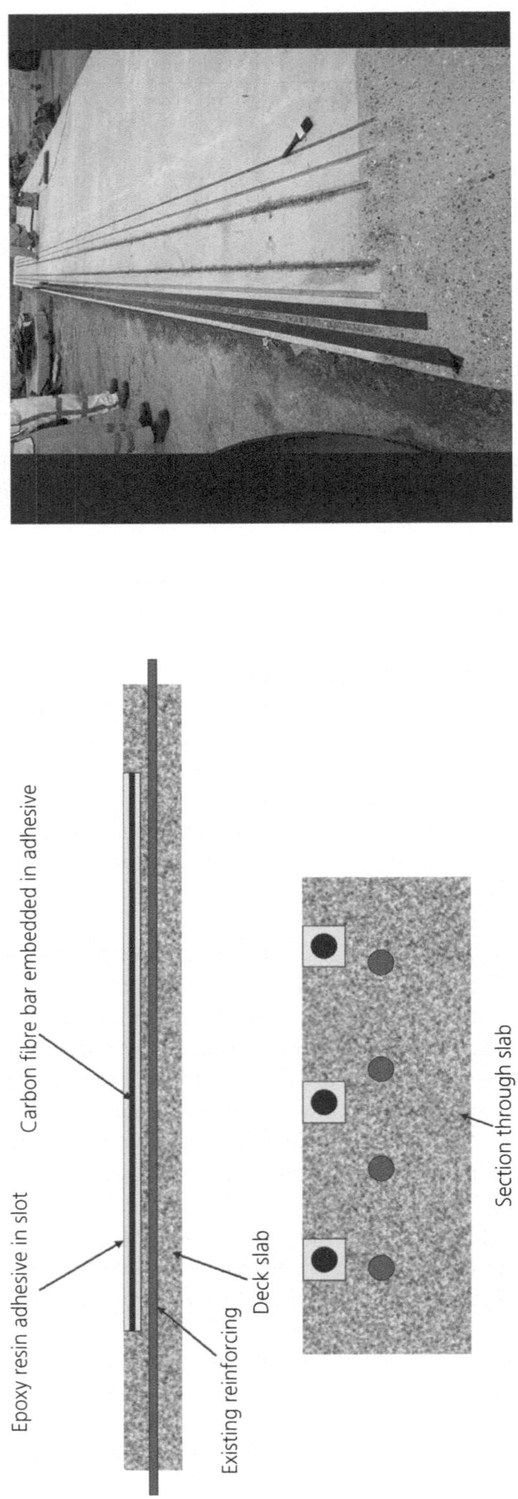

Figure 9.1 Illustration of the NSM system and the interaction with steel reinforcement (Image courtesy of Concrete Repairs Limited UK)

Figure 9.2 Symbols describing geometry of the NSM slot and clearances (Reproduced from *BS EN 1992-1-1:2023* and amended)

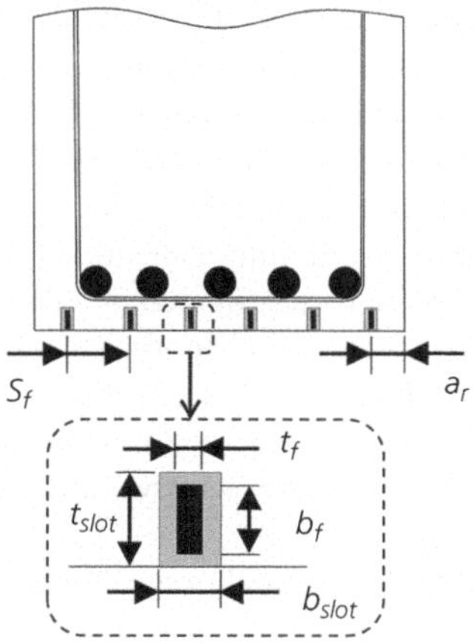

Table 9.1 NSM geometric limitations for slots, edge distances and spacing (Reproduced from *BS EN 1992-1-1:2023*)

Geometrical limit	Square NSM bars	Round NSM bars	NSM strips
Slot width	$t_f + 2 \leq b_{slot} \leq t_f + 6$	$Ø_f + 2 \leq b_{slot} \leq Ø_f + 6$	$t_f + 2 \leq b_{slot} \leq t_f + 4$
Slot depth	$b_f + 1 \leq t_{slot} \leq b_f + 3$ $t_f \leq b_f$	$Ø_f + 1 \leq t_{slot} \leq Ø_f + 3$	$b_f \leq t_{slot} \leq b_f + 2$
Edge distance	$a_r \geq 4b_f$	$a_r \geq 4b_f$	$a_r \geq 4b_f$
Centre-to-centre spacing of NSM	$max\{3b_{slot}, D_{upper}\}$ $\leq s_f \leq$ $(min\{0.2l_{0b}, 3h\})$*	$max\{3b_{slot}, D_{upper}\}$ $\leq s_f \leq$ $(min\{0.2l_{0b}, 3h\})$*	$max\{b_f, D_{upper}\}$ $\leq s_f \leq$ $(min\{0.2l_{0b}, 3h\})$*

*Denotes that the condition applies only to beams in flexure
Where
All dimensions are in mm
l_{0b} is the distance between adjacent points of contraflexure in which the strengthening system is placed
h is the overall depth of the member
$ø_f$ is the diameter of an NSM bar or rod.

for concrete cover separation across the width of the beam and force the assumed failure mechanism of debonding around the concrete slot perimeter.

- The centre-to-centre spacing limit also prevents overlapping of confining stress zones around each NSM bar, maximising bond resistance of the system, as shown in Hassan and Rizkalla (2004) and Rashid *et al.* (2008).
- The minimum edge distance required ($\geq$ 4 times the depth of the slot) is intended to avoid the separation of concrete on the corner of a member in bending, also as shown in Hassan and Rizkalla (2004) and Rashid *et al.* (2008).

Strengthening with EBR strips and sheets

Flexural strengthening

Determining the position of termination points for EBR CFRP systems has been dealt with in Chapter 3. To recap, in order to comply with *BS EN 1992-1-1:2023*, a CFRP strengthening system must extend beyond the region where strengthening is needed, to a point where any tensile force that can be generated in the system is resisted by the design bond strength of the anchorage, F_{bfRd}. The anchorage needs to be assessed for the potential for end cover separation and, if the requirements of the code are not met, additional CFRP shear strengthening needs to be applied to prevent this occurring. Where shear strengthening is needed the shear stirrup closest to end of the longitudinal CFRP strengthening needs to be terminated at a distance not greater than 50 mm from the end of the CFRP strengthening, as shown in Figure 3.8. The anchorage area also needs to be checked for the potential for shear induced crack separation and, if the requirements of the code are not met, additional CFRP shear strengthening again needs to be applied to prevent this occurring. Shear induced crack separation should be assessed throughout the length of strengthening.

When determining the force in the CFRP at anchorage, the shift term in *BS EN 1992-1-1:2023* needs to be accounted for.

BS EN 1992-1-1:2023, J.12.1 provides limitations on the maximum spacing of individual strips, $s_{f,max}$, in an EBR flexural strengthening system. These are reproduced below, and slightly reformatted for ease of use. The provisions are used to ensure that large sections of the surface being strengthened do not remain unreinforced with CFRP.

For a cantilevered member in flexure Equation 9.1 applies

$$s_{f,max} = min(0.4l, 400 \text{ mm}) \tag{9.1}$$

Where l is the length of the cantilever.

For any other member in flexure Equation 9.2 applies

$$s_{f,max} = min(0.2l_{0b}, 400 \text{ mm}) \tag{9.2}$$

The clause also requires that the distance from the outer edge of the outermost EBR strip or sheet to the edge of the member is not greater than the nominal cover, c_{nom}. This similarly avoids the strengthening becoming ineffective at the edge of a member strengthened with an EBR CFRP system subjected to flexure.

Multi-layered systems

BS EN 1992-1-1:2023, J.12.4 states the following with regard to multi-layered systems.

- No more than five layers of CF sheets should be used for flexural strengthening.
- No more than ten layers of CF sheets should be used for confinement strengthening of a column.
- No more than two layers of EBR strips should be used for any type of strengthening.
- The maximum thickness of EBR strengthening should not exceed 3 mm, excluding the adhesive.

BS EN 1992-1-1:2023 is silent on how layered systems should be detailed at termination points. TR55 (Concrete Society, 2013), ACI 440.2R-17 (ACI, 2017) and fib Bulletin No. 90 (fib, 2019) all recommend a staggering of the end of individual layers of strengthening to ensure that stress concentrations do not cause effects not considered in design. fib Bulletin No. 90 and ACI 440.2R-17 are consistent in recommending an overlap of at least 150 mm. Figure 9.3 provides guidance on how this could be practically achieved.

Orthogonal systems

A member can be strengthened in two orthogonal directions on the same surface with EBR strips or sheets. When taking into account the thickness of the layer closest to the concrete surface, the outermost layer will need a change in thickness of the system at the intersection. If not implemented correctly, tensile stresses perpendicular to the strengthened surface can be produced, which are not accounted for in the code, increasing the risk of the system debonding at this crossing point. fib Bulletin No. 90 (fib, 2019) recommends that the unevenness rules for strips and sheets described in Chapter 10 are complied with in this instance. Given that this is quite common, it is recommended that the supplier is engaged with to ensure that detailing is undertaken in a manner that does not increase the risk of debonding of the system.

Shear strengthening

With the absence of any specific guidance in *BS EN 1992-1-1:2023* pertaining to spacing limits of shear strengthening, the provisions of TR55 (Concrete Society, 2013) present a pragmatic approach to ensuring that the CFRP will cross shear cracks, as assumed in the provisions in the code. The provisions of TR55 are reproduced using the terminology in *BS EN 1992-1-1:2023* in Equations 9.3 and 9.4 for ease.

$$s_f = min\begin{cases} 0.8d_f \\ d_{fv} - l_{bf,max,k}\dfrac{n_s}{3}cos\beta \\ b_f + \dfrac{d_f}{4} \end{cases} \tag{9.3}$$

Where

$$d_{fv} = h_f - \left(h - d_{st}\right) \tag{9.4}$$

h is the overall depth of the member in flexure

h_f is the height of CFRP shear strengthening indicated in Figure 4.2

d_{st} is the effective depth of the cross section

$$n_s = \begin{cases} 0 \text{ for a fully wrapped beam} \\ 1 \text{ for an open } \left(\text{U wrapped}\right) \text{ strengthening systems} \end{cases}$$

and $\beta = 90° - \alpha_f$ (in degrees)

Anchorage of fully wrapped systems

Anchorage requirements for fully lapped systems are not codified in a manner that enables the designer to specify a length. This is because they are intrinsically linked to the combination of CFRP product and adhesive used by a supplier. This must be a length specified by the supplier, and stated on any drawings, models or specifications. TR55 (Concrete Society, 2013) does, though, recommend that such laps should not be less than 200 mm which could be used as guidance.

Where multiple layers of CFRP are used for confining a column, laps should be staggered around the column perimeter, an example of which is shown in Figure 9.4.

Figure 9.3 Termination of multi-layered EBR CFRP

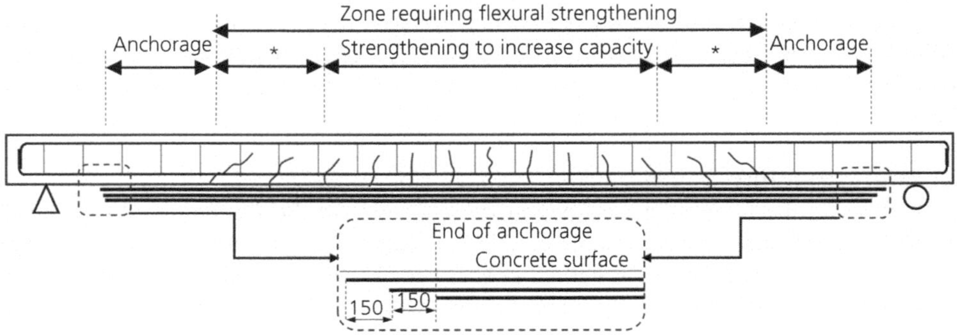

Notes

- * Denotes that the CFRP will extend through this area until the design anchorage resistance can sustain forces generated from compatibility
- Dimensions are in mm where shown

Figure 9.4 Wrapping a concrete column with multiple plies

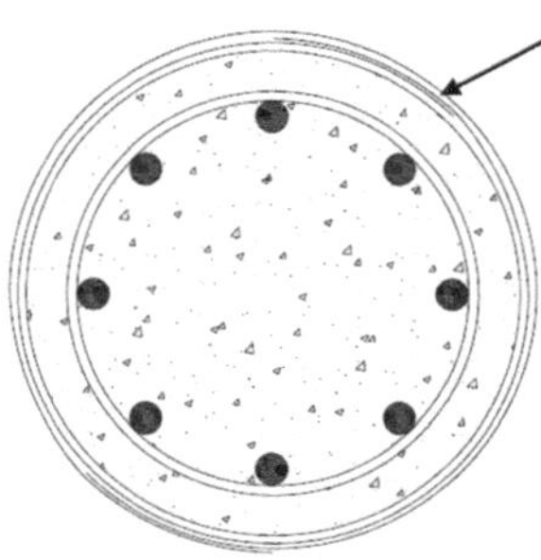

Laps staggered as shown

Bending of prefabricated CFRP elements

BS EN 1992-1-1:2023 includes the following limits for bending of prefabricated elements, including EBR strips and bars, and NSM bars and rods.

- Straight strips and bars should not be designed to curve more than 1000 times their thickness when installed, unless bending effects from curvature are explicitly taken into account in the design.
- If a curved system is fabricated, the manufacturer will need to ensure that the value for the characteristic short-term tensile strength, f_{fuk}, provided accounts for the curve and bending effects that may arise from the curvature of the installed system due to deformation of the structure.
- As stated in *BS EN 1992-1-1:2023*, J.11.1.2.3, if action effects result in a curvature which produces a component of tension between the CFRP system and the concrete surface onto which it is bonded, the bond strength will be reduced, and specialist advice will need to be obtained.
- Where NSM elements are curved, the radius should be greater than 2 m.

Kaloyana Kostova and Craig Giaccio
ISBN 978-1-83549-935-1
https://doi.org/10.1108/978-1-83549-934-420251010

Chapter 10
Considerations for execution

Introduction

Installation of CFRP systems on site is a specialist activity. This chapter does not intend to provide a detailed specification for the manufacture and application of the system itself, but instead intends to inform designers of matters that are relevant to design and may impact design decisions. The basic premise in this chapter has been to identify works required to the concrete surface, as this is generally within the remit of the designer, particularly in the early stages of evaluating the feasibility of a CFRP strengthening scheme.

Designers should also consult BS EN 1504-10:2017 (BSI, 2017), as it provides requirements for the matters listed below, most of which are not discussed in this book

- substrate condition before and during application of systems and products
- storage of systems and products
- structural stability during preparation, protection and repair
- method of protection and repair
- quality control for execution of the work
- maintenance of the structure.

Adequacy of substrate for strengthening

The substrate must be free of deterioration at the time strengthening is applied. *BS EN 1992-1-1:2023* (BSI, 2023b), Annex I provides an indication of what deterioration mechanisms are prevalent in aging concrete structures. Publications such as TR54 (Concrete Society, 2000) provide good guidance on how to diagnose these deterioration mechanisms.

It is important to realise that it is not only defects visible to the naked eye that can lead to a strengthening system not realising its design resistance – for example, delamination caused by reinforcement corrosion can occur prior to signs of corrosion being visible on the surface of the concrete. To ensure that all deterioration mechanisms are identified, it is recommended (similarly to TR55 (Concrete Society, 2013)) that a thorough visual inspection, assessment of the surface tensile strength, a sounding survey and chemical analyses be undertaken to confirm the adequacy of the substrate for strengthening.

If defects are found, which need to be remediated, and strengthening with CFRP is pursued, remedial works will need to be mindful of compatibility with the CFRP strengthening system. For example, if epoxy resin injection is used to resolve honeycombing that may be present near the surface of a member being strengthened, it should be compatible with the CFRP system adhesive and/or resin if there is any chance that the epoxy resin can come into contact with any of them.

Figure 10.1 Removal of deteriorated concrete pre-strengthening (Photo courtesy of Concrete Repairs Limited UK)

An illustration of concrete removal that may be undertaken prior to strengthening is shown in Figure 10.1.

As stated in Chapter 9, the presence of a curvature, opposite to that which causes tension on the surface of a member being strengthened in flexure, can lead to transverse stresses on the CFRP resulting in the CFRP peeling away. The inspection for strengthening should evaluate the presence of such a curvature. If it is present, appropriate action should be taken, which can vary from removing the friction term from the intermediate crack debonding check through to obtaining expert advice as to how to deal with the specific issue.

Surface preparation and installation considerations for EBR systems

Once a substrate is confirmed as adequate to receive strengthening, the next step is to prepare the surface such that it will comply with the assumptions inherent in the design provisions contained in the standard.

A CFRP strengthening system generally needs to be applied to a dry and clean surface, free of defects. Where not possible (e.g. in an underwater application), a supplier will need to be engaged early in the design process to ensure that all relevant considerations are taken into account during the design development process.

From the perspective of the bond checks that are undertaken in the code, the most important aspect for the existing structure is the quality of the concrete onto which the CFRP will be adhered.

Cement laitance must be removed from the surface to expose aggregate in the concrete and ensure that all loose material is removed. Al-Mahaidi and Kalfat (2018) recommend leaving a surface roughness of 0.6–1.0 mm, similar to that of 60-grit sandpaper. The surface must be dry and clean (including removal of oil and grease) prior to installation of the CFRP strengthening system.

The method selected to roughen the substrate should not be harsh enough to damage it, or even cause micro-cracking. This generally precludes mechanical and pneumatic methods, leaving grit blasting, water jetting, sand blasting or grinding as the most suitable alternatives.

Cracks can cause debonding. Therefore, significant pre-existing cracks should be injected prior to installation of the CFRP so that debonding effects not considered in the design checks do not arise. This will generally be done with an epoxy resin. Selection of this resin will need to be done in conjunction with the CFRP supplier as the selected resin will need to be compatible with their strengthening system. Injection or surface filling will also be needed if voids (e.g. from honeycombing) or bug holes are present.

Where CFRP needs to bend through a corner, rounding of the corners is needed to avoid damage to the CFRP. This is also a key consideration for shear and confinement strengthening as the corner radius is a key parameter in determining the design shear strength of a CFRP system, f_{fwd}, and the increase in compressive strength from confining with CFRP, Δf_{cd}. Minimum radii of around 15 mm are specified in ACI 440.2R-17 (ACI, 2017) and TR55 (Concrete Society, 2013). fib Bulletin No. 90 (fib, 2019) suggests 30 mm, when corner rounding is required. Given the discrepancy in advice in published literature, it is advisable to confirm the minimum requirement for a given system with the supplier where possible. It is also advisable to ensure that the radius is as large as possible to ensure that the maximum confining force for the CFRP system can be achieved.

The surface on which the CFRP will be bonded will also need to be made even. This is to ensure that localised transverse stresses caused by straightening of the CFRP, which can cause tension in the adhesive, are eliminated. As stated in Chapter 3, ACI 440.2R-17 (ACI, 2017) recommends that the surface curvature of concave surfaces should not exceed 5 mm per metre, and TR55 (Concrete Society, 2013) recommends the same for plate-based systems and 3 mm per metre for CF sheets. fib Bulletin No. 90 (fib, 2019) provides a bit more detail on this topic, and recommends the unevenness criteria summarised in Table 10.1.

Table 10.1 Guidance on surface unevenness for installation of CFRP strengthening, taken from fib Bulletin No. 90 (Table copied from fib Bulletin No. 90)

Type of CRP	Permissible deviation over a 2m long surface*	Permissible deviation over a 0.3 m long surface*
Pre-cured systems with thickness > 1mm	10	4
Pre-cured systems with thickness < 1mm	6	2
Systems that are cured in situ	4	2

*Denotes that unevenness is measured as the maximum distance between the concrete surface and the surface formed by the strengthening system at the concrete-to-CFRP interface

Once the substrate is prepared, the substrate strength will need to be checked in accordance with BS EN 1542:1999 (BSI, 1999). fib Bulletin No. 90 (fib, 2019) recommends that a minimum surface tensile strength of 1.5 MPa is observed on site. Tests to prove this will enable the designer to validate the surface tensile strength of concrete, $f_{ctm,surf}$, assumed in design, prior to installation of the CFRP system. When the provisions in this code are used by a designer producing a detailed design ahead of procuring the CFRP supplier, it is recommended that a requirement to validate the assumed design value of $f_{ctm,surf}$ should be clearly articulated in the project specification produced by the designer for construction.

As mentioned in Chapter 2, the glass transition temperature, T_g, limits the performance of the CFRP system. This could become an issue in situations such as application of hot rolled asphalt onto the top surface of a bridge deck which has been strengthened with CFRP (e.g. in hogging). In such instances, protection using cover systems that ensure the temperature of the CFRP does not reach T_g may be required.

The performance of adhesives, including curing times, can be influenced by the temperature at installation. fib Bulletin No. 90 (fib, 2019) recommends that application should be undertaken where the surface temperature is above 3 °C. Where the temperature is lower than 3 °C, dehumidifiers and surface heating may be required, and these would depend on the supplier's system. Al-Mahaidi and Kalfat (2018) cite challenges in installing FRP strengthening systems in temperatures greater than 35 °C because of reduced working times at such elevated temperatures and difficulty in applying adhesive to the structure. While it is not clear from this reference, one cannot rule out that the difficulty in application can potentially arise if the surface temperature of a structural member is greater than 35 °C. Refrigeration may be needed in such elevated temperatures. It is recommended that air and surface temperature ranges are referenced in specification for execution documents.

Where the EBR system is applied for confining columns, ACI 440.2R-17 (ACI, 2017) guidelines for execution can be used. In summary, these note that care needs to be taken to ensure full contact around the perimeter of the member and for the full height. Large voids need to be repaired with a material compatible with the concrete and finished with a profile that matches the profile of the existing member. Surface finishes (e.g. plaster) need to be removed to get back to bare concrete, so that adequate contact can be made for strengthening.

Surface preparation for NSM systems

NSM systems should only be used where there is sufficient cover to the lowest layer of reinforcement (including stirrups) to avoid damage when slots are cut into the concrete in which rods or bars are placed. The impact of reductions in cover from cutting slots into the concrete, and replacement with the NSM CFRP and epoxy, on the residual life also needs to be considered. A designer can choose to position NSM between bars in the bottom mat of reinforcement to avoid cutting bars. A cover meter survey should be undertaken as early as possible during the design phase to ensure that existing reinforcement or prestressing does not interfere with the intended position of the system.

Once recesses are cut into the structure, they need to be prepared. Given that the slots cut through the laitance and looser surface materials, less preparation is needed at this point than for EBR systems. They simply need to be cleaned and laitance-free to ensure adequate bond is achieved with the adhesive.

The guidance provided in this chapter for adequacy of a substrate for strengthening with EBR systems applies equally to NSM systems as well. At the point at which the grooves have been cut and are ready to receive the strengthening system, the surface should be rechecked for adequacy, in particular a check for voiding or honeycombing near to the new surface (albeit a small surface) that has been created. If any defect is detected at this point, it will need to be remedied using materials that are compatible with the strengthening system.

The following items described in the section on surface preparation for EBR systems in this chapter apply equally to NSM systems

- temperature at installation of the system
- injection of cracks may be needed to avoid delamination of the CFRP system.

Fire performance

Design of strengthening with CFRP for fire resistance was not introduced into *BS EN 1992-1-1:2023*, Annex J, and was not considered in *BS EN 1992-1-1:2023*. Design considerations for fire are discussed in the standard, as outlined in Chapter 5. It is noted that products used to enhance fire performance are available from CFRP suppliers. Should this be a consideration for execution, advice should be sought from a supplier.

Kaloyana Kostova and Craig Giaccio
ISBN 978-1-83549-935-1
https://doi.org/10.1108/978-1-83549-934-420251011

Chapter 11
References

ACI (American Concrete Institute) (2017) ACI 440.2R-17 Guide for the design and construction of externally bonded FRP systems for strengthening concrete structures. ACI, Michigan, USA.

ACI (American Concrete Institute) (2019) ACI 318-19 Building code requirements for structural concrete (ACI 318-19) and commentary (ACI 318R-19). ACI, Michigan, USA.

Al-Mahaidi R and Kalfat R (2018) *Rehabilitation of Concrete Structures with Fibre-Reinforced Polymer*, 1st edn. Elsevier.

AS (2017) AS 5100.8:2017 Bridge design rehabilitation and strengthening of existing bridges. Standards Australia (AS).

Binici B (2008) Design of FRPs in circular bridge column retrofits for ductility enhancement. *Engineering Structures* **30(3)**: 766–776.

BSI (British Standards Institution) (1996) BS EN 2561:1996 Aeronautical – plastics reinforced with carbon fibres – unidirectional laminates – tensile test parallel to the fibre direction. BSI, London, UK.

BSI (British Standards Institution) (1999) BS EN 1542:1999 Products and systems for the protection and repair of concrete structures. Test methods. Measurement of bond strength by pull-off. BSI, London, UK.

BSI (British Standards Institution) (2003) BS EN 1991-1-5:2003 Eurocode 2. Eurocode 1. Actions on structures – general actions – thermal actions. BSI, London, UK.

BSI (British Standards Institution) (2004a) BS EN 1992-1-1:2004 Eurocode 2. Design of concrete structures – Part 1-1: General rules and rules for buildings. BSI, London, UK.

BSI (British Standards Institution) (2004b) BS EN 1504-4:2004 Products and systems for the protection and repair of concrete structures. Definitions, requirements, quality control and evaluation of conformity – Structural bonding. BSI, London, UK.

BSI (British Standards Institution) (2004c) BS EN 12614:2004 Products and systems for the protection and repair of concrete structures – test methods – determination of glass transition temperatures of polymers. BSI, London, UK.

BSI (British Standards Institution) (2004d) BS EN 1998-1:2004+A1:2013 Eurocode 8. Design of structures for earthquake resistance – General rules, seismic actions and rules for buildings. BSI, London, UK.

BSI (British Standards Institution) (2017) BS EN 1504-10:2017 Products and systems for the protection and repair of concrete structures – definitions, requirements, quality control and evaluation of conformity – Part 10: Site application of products and systems and quality control of the works. BSI, London, UK.

BSI (British Standards Institution) (2023a) BS EN 1990:2023 Eurocode. Basis of structural and geotechnical design. BSI, London, UK.

BSI (British Standards Institution) (2023b) BS EN 1992-1-1:2023 Eurocode 2. Design of concrete structures. General rules and rules for buildings, bridges and civil engineering structures. BSI, London, UK.

Concrete Society (2000) *Technical Report No. 54 Diagnosis of Deterioration in Concrete Structures: Identification of Defects, Evaluation and Development of Remedial Action*, 1st edn. The Concrete Society, Berkshire, UK.

Concrete Society (2013) *Technical Report No. 55 Design Guidance for Strengthening Concrete Structures Using Fibre Composite Materials*, 2nd edn. The Concrete Society, Berkshire, UK.

Darby AP, Ibell TJ, Evernden MC and Coonan R (2011) FRP confined square and rectangular columns under concentric and eccentric loading. *Proceedings of the 5th International Conference Advanced Composites in Construction*: 264–275.

EOTA (European Technical Approval) (2019) EAD 160086-00-0301 – Kits for the strengthening of concrete elements by externally bonded CFRP strips. See https://www.eota.eu/download?file=/2017/17-16-0086/for%20ojeu/ead%20160086-00-0301_ojeu%202023.pdf (accessed 22/12/2024).

European Committee for Standardisation (2023) Background Document for FprEN 1992-1-1, CEN/TC 250/SC 2 N2087. CEN, Brussels, Belgium. (unpublished)

fib (2019) fib Bulletin No. 90, *Externally Applied FRP Reinforcement for Concrete Structures*. The International Federation for Structural Concrete [fib], Germany.

Giaccio C (2003) *Flange Effectiveness in the Resistance of Shear on RC T-Beams Subjected to Point Loads*. PhD Thesis, University of Adelaide, Adelaide, Australia.

Hassan T and Rizkalla S (2004) Bond mechanism of near-surface-mounted fiber-reinforced polymer bars for flexural strengthening of concrete structures. *ACI Structural Journal* **101(6)**: 830–839.

Hendy CR, Smith DA and Gulvanessian H (2007) *Designers' Guide to EN 1992-2. Eurocode 2: Design of Concrete Structures. Part 2: Concrete Bridges*, 1st edn. Thomas Telford, London, UK.

ISO (International Organization for Standardization) (1997) ISO 14130:1997 Fibre-reinforced plastic composites – determination of apparent interlaminar shear strength by short-beam method. ISO, Geneva, Switzerland.

ISO (International Organization for Standardization) (2013) ISO 14484:2013 Performance guidelines for design of concrete structures using fibre-reinforced polymer (FRP) materials. ISO, Geneva, Switzerland.

ISO (International Organization for Standardization) (2015a) ISO 10406-1:2015 Fibre-reinforced polymer (FRP) reinforcement of concrete – Test methods – Part 1: FRP bars and grids. ISO, Geneva, Switzerland.

ISO (International Organization for Standardization) (2015b) ISO 10406-2:2015 Fibre-reinforced polymer (FRP) reinforcement of concrete – Test methods – Part 2: FRP sheets. ISO, Geneva, Switzerland.

ISO (International Organization for Standardization) (2019a) ISO 527:2019 Plastics – Determination of tensile properties – Part 1: General principles. ISO, Geneva, Switzerland.

ISO (International Organization for Standardization) (2019b) ISO 10406-3:2019 Fibre-reinforced polymer (FRP) reinforcement of concrete – Test methods – Part 3: CFRP strips. ISO, Geneva, Switzerland.

Kalupahana WK, Ibell T and Darby A (2013) Bond characteristics of near surface mounted CFRP bars. *Construction and Building Materials* **43**: 58–68. https://doi.org/10.1016/j.conbuildmat.2013.01.021

Lam L and Teng JG (2003a) Design-oriented stress-strain model for FRP-confined concrete. *Construction and Building Materials* **17**: 471–489.

Lam L and Teng JG (2003b) Design-oriented stress-strain model for FRP-confined concrete in rectangular columns. *Journal of Reinforced Plastics and Composites* **22(13)**: 1149–1186.

Muttoni A and Fernández Ruiz M (2008) Shear strength of members without transverse reinforcement as function of critical shear crack width. *ACI Structural Journal* **105(2)**: 163–172.

Rashid R, Oehlers DJ and Seracino R (2008) IC debonding of FRP NSM and EB retrofitted concrete: plate and cover interaction tests. *Journal of Composite Construction* **12(2)**: 160–167.

Tajaddini A, Ibell T, Darby A, Evernden M and Silva P (2016) Quantifying moment redistribution in FRP-strengthened RC beams. *Proceedings of the Institution of Civil Engineers: Structures and Buildings* **169(11)**: 853–862, 10.1680/jstbu.15.00126.

Kaloyana Kostova and Craig Giaccio
ISBN 978-1-83549-935-1
https://doi.org/10.1108/978-1-83549-934-420251012

Annex A
Flow chart for flexural design of a member

The iterative procedure described herein may be used for calculating capacity of non-uniform cross sections which are strengthened and unstrengthened, as well as determining strain profiles under externally applied loads. The main difference with the standard stress-strain analysis is the inclusion of two different reinforcement types and the pre-strain in concrete.

The stress-strain analysis of sections strengthened with CFRP reinforcement should follow the steps illustrated in the flow chart in Figure 12.1 to obtain the strengthened capacities and strains under ultimate limit state (ULS) loads. The force in CFRP can then be determined, from the strains at different sections along the length of a member, to carry out the anchorage and intermediate crack debonding checks.

The algorithms for calculating the pre-strain, ε_0, the strengthened capacity in the case of concrete crushing, the strengthened capacity in the case of CFRP rupture and the strain profile after strengthening are based on a layer-by-layer approach used to obtain the forces in the concrete, as illustrated in Figure 12.2. The stresses in the concrete and the reinforcements are derived from the stress-strain relationships, defined in *BS EN 1992-1-1:2023* (BSI, 2023b). The following assumptions apply.

- Strains are assumed to be linearly distributed along the cross-sectional depth.
- Perfect bond is assumed between the steel reinforcing bars and the surrounding concrete.
- Perfect bond is assumed between the CFRP reinforcement and the surrounding concrete.

A numerical root-finding method is then used to find the neutral axis from the equilibrium of all forces acting on the cross section under consideration (*Find_x*) and to find the strain in the extreme concrete fibre in compression (*Find_ε*) from the equilibrium of moments under applied loads. In the program written to obtain the results presented in this book, a simple binary search (also known as a half-interval search) was employed but other root-finding methods may reduce the computational time.

The number of layers depends on the chosen *step* (depth of a layer) and can be varied to achieve the desired accuracy. The application of the method for finding the pre-strain in concrete is summarised in Figure 12.3. The width of the concrete section is defined as a function of the section depth to allow for non-rectangular geometries. The steel reinforcement can also be introduced for a number of layers n_{reinf}. The equation for finding the top concrete strain is solved iteratively by finding, again iteratively, the neutral axis for each iteration. Once the concrete strain and neutral axis are obtained, the pre-strain in concrete and steel can be calculated at every layer.

Figure 12.1 General flow chart for analysis of CFRP strengthened sections

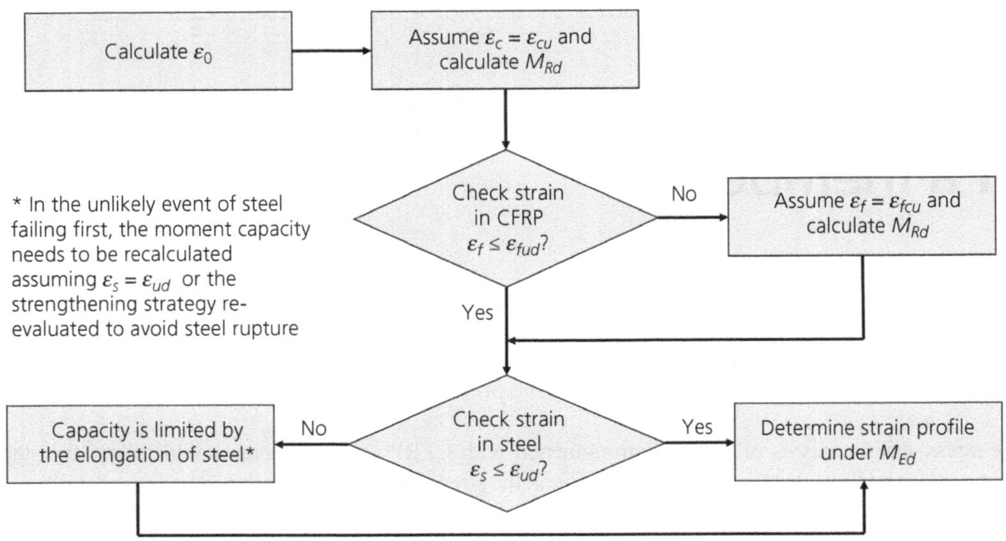

Figure 12.2 Layer-by-layer calculation of forces in concrete

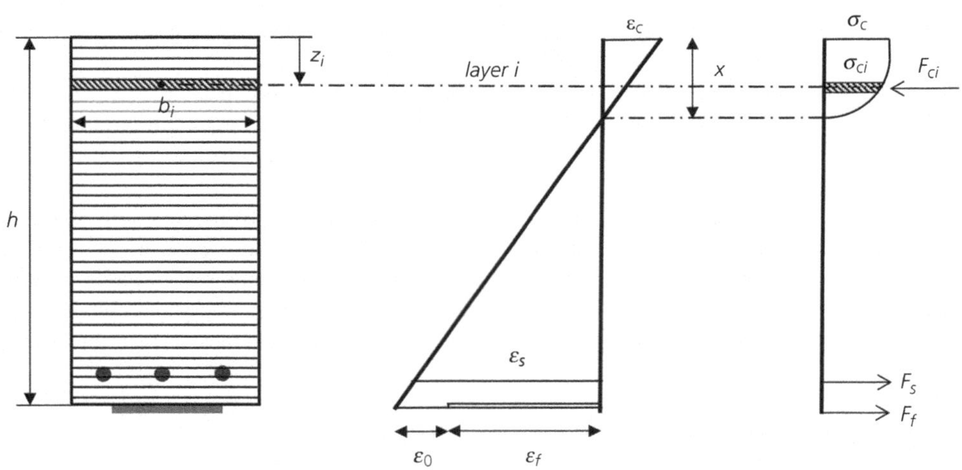

The strain distribution after strengthening is indicated in Figure 12.4. The total compressive strain in concrete is equal to the top pre-strain plus the top strain due to the additional moment. This also applies to the strains in steel reinforcement, while the strain in CFRP is generated only by the additional moment. Figure 12.5 illustrates how this can be accounted for in the algorithm for calculating the strengthened capacity in the case of concrete crushing in compression first.

Figure 12.3 Flow chart for finding pre-strain in concrete

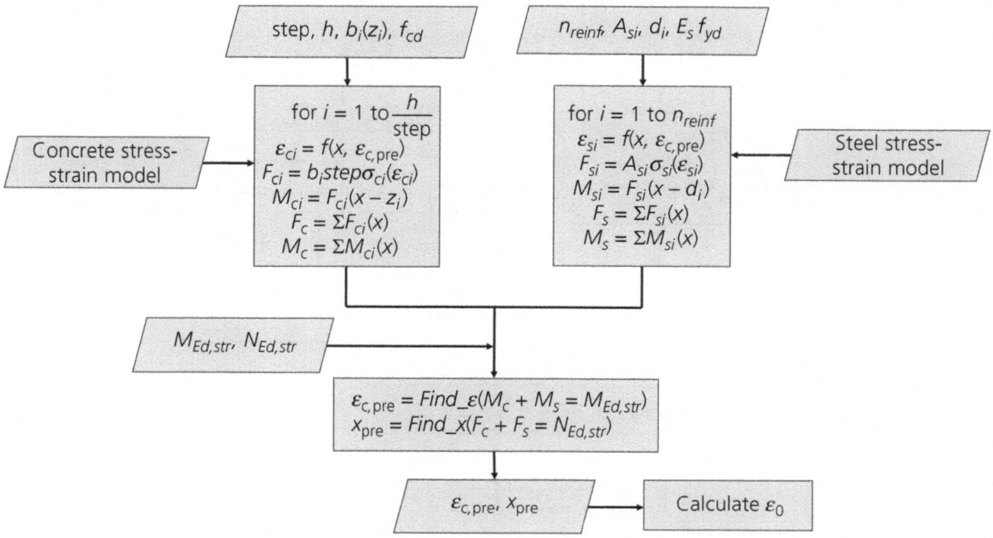

Figure 12.4 Strain profiles before and after strengthening including the additional moment

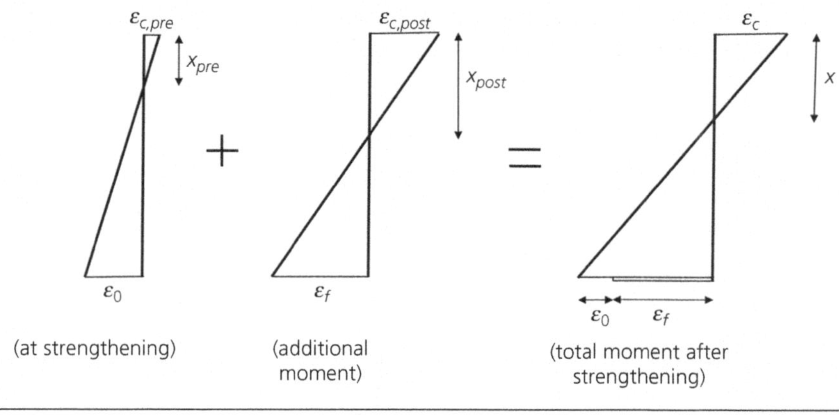

In order to calculate the strengthened capacity corresponding to CFRP rupture, the top strain in concrete can be expressed as a function of the neutral axis and the ultimate strain in CFRP, as shown in Equation 12.1.

$$\varepsilon_c = \frac{x}{d_f - x}\left(\varepsilon_{fud} + \varepsilon_0\right) \tag{12.1}$$

Figure 12.5 Flow chart for finding strengthened moment capacity when concrete fails in compression first

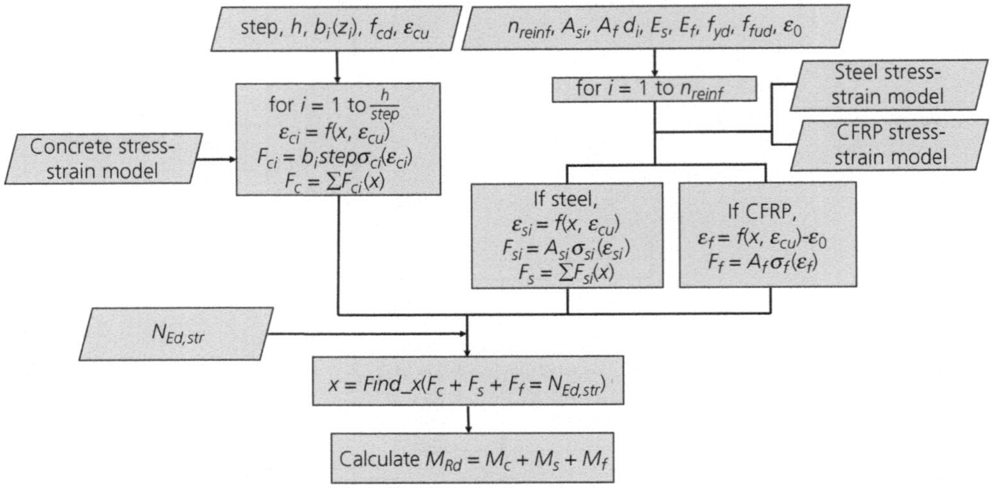

The described procedure is also appropriate for general stress-strain analysis under ULS load by solving the two equilibrium equations, similarly to finding the pre-strain in Figure 12.3. The procedure can be applied to any number of sections along the length of a strengthened member to produce CFRP stress and force diagrams, and to calculate force differences. Full examples are provided in Annex B.

The relationships in Annex D and E can be applied to avoid using layered sections where the compression region of the strengthened member is rectangular.

Kaloyana Kostova and Craig Giaccio
ISBN 978-1-83549-935-1
https://doi.org/10.1108/978-1-83549-934-420251013

Annex B
Ultimate limit states design examples

Additional examples of flexural strengthening are provided in this annex to illustrate various aspects of the code requirements. They all use the iterative strain-based method described in Annex A to find moment resistances and forces to enable debonding checks to be undertaken.

Example Flexural Design B-1 (EBR CF sheets):

Calculate the capacity of the simply supported T-beam illustrated in Figure 13.1, assuming the beam is strengthened in flexure using wet lay-up CF sheets under the given loading configuration.

Reinforcement details

- Tensile reinforcement: 10 numbers 32 mm-diameter bars in two layers
- Compression reinforcement: 2 numbers 12 mm-diameter bars
- Shear links: 16 mm-diameter shear links at 200 mm spacing
- Concrete cover to stirrups: 50 mm

Assume spacing of 1 bar diameter between the tensile reinforcement layers. Therefore, the effective depths and steel areas are

- Bottom layer 1 (tension): $d_{st.1} = 1143\,\text{mm}\ A_{st.1} = 4020\,\text{mm}^2$
- Bottom layer 2 (tension): $d_{st.2} = 1079\,\text{mm}\ A_{st.2} = 4\,020\,\text{mm}^2$
- Top layer (compression): $d_{sc.1} = 72\,\text{mm}\ A_{sc.3} = 226\,\text{mm}^2$
- Shear links: $A_{sw} = 402\ \text{mm}^2$

Figure 13.1 T-beam cross section and loading configuration – all dimensions are in mm unless stated otherwise

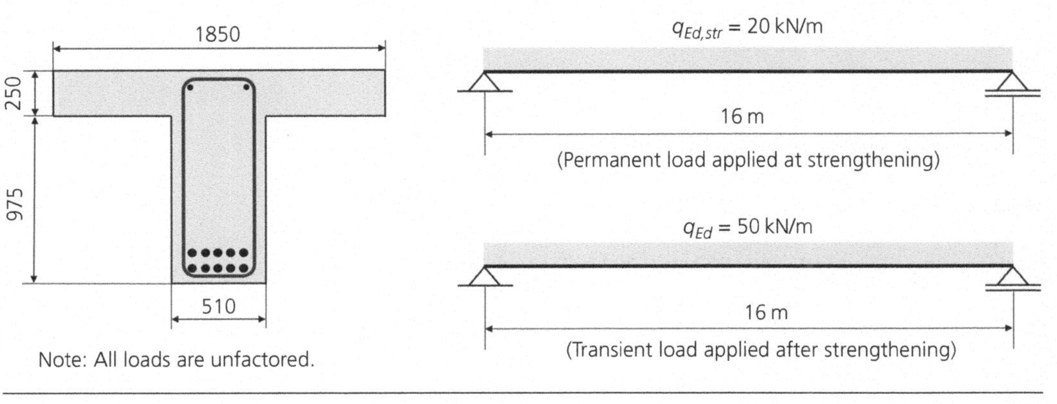

Materials

Concrete

Characteristic compressive strength: $f_{ck} = 30$ MPa

Mean compressive strength (*BS EN 1992-1-1:2023* (BSI, 2023b), Table 5.1): $f_{cm} = 38$ MPa

Mean tensile strength (*BS EN 1992-1-1:2023*, Table 5.1): $f_{ctm} = 2.9$ MPa

Modulus of elasticity (*BS EN 1992-1-1:2023*, Formula 5.1):

$$E_{cm} = K_E f_{cm}^{\frac{1}{3}} = 9500 \times 38^{\frac{1}{3}} = 31939\,\text{MPa}$$

Partial safety factor (*BS EN 1992-1-1:2023*, Table 4.3): $\gamma_c = 1.5$

Design compressive strength (*BS EN 1992-1-1:2023*, Formula 5.3):

$$f_{cd} = \frac{\eta_{cc} k_{tc} f_{ck}}{\gamma_c}, \text{ where } \eta_{cc} = \left(\frac{f_{ck,ref}}{f_{ck}}\right)^{\frac{1}{3}} \leq 1.0$$

$$\eta_{cc} = \left(\frac{f_{ck,ref}}{f_{ck}}\right)^{\frac{1}{3}} = \left(\frac{40}{30}\right)^{\frac{1}{3}} = 1.1 > 1.0$$

Take $k_{tc} = 0.85$ (*BS EN 1992-1-1:2023*, 5.1.6 (1) NOTE)

$$f_{cd} = \frac{1 \times 0.85 \times 30}{1.5} = 17 \text{ MPa}$$

Design tensile strength (*BS EN 1992-1-1:2023*, Formula 5.5)

$$f_{ctd} = \frac{k_{tt} f_{ctk,0.05}}{\gamma_c}$$

Characteristic tensile strength 5%-fractile (*BS EN 1992-1-1:2023*, Table 5.1): $f_{ctk,0.05} = 2.0$ MPa

Take $k_{tt} = 0.7$ (*BS EN 1992-1-1:2023*, 5.1.6 (2) NOTE)

$$f_{ctd} = \frac{1 \times 0.7 \times 2.0}{1.5} = 0.9 \text{ MPa}$$

Steel reinforcement

Characteristic tensile strength: $f_{yk} = 400$ MPa

Modulus of elasticity: $E_s = 200\,000\,\text{MPa}$

Partial safety factor (*BS EN 1992-1-1:2023*, Table 4.3): $\gamma_s = 1.15$

Design tensile strength (*BS EN 1992-1-1:2023*, Formula 5.11): $f_{yd} = \dfrac{f_{yk}}{\gamma_s} = \dfrac{400}{1.15} = 348$ MPa

FRP strengthening properties

Characteristic short-term tensile strength: $f_{fuk} = 3850$ MPa

Mean modulus of elasticity: $E_f = 225\,000$ MPa

Characteristic ultimate strain at rupture: $\varepsilon_{fuk} = \frac{f_{fuk}}{E_f} = \frac{3850}{225\,000} = 0.0171 = 1.71\%$

Design tensile strength (*BS EN 1992-1-1:2023*, Formula J.2): $f_{fud} = \frac{\eta_f f_{fuk}}{\gamma_f}$

Partial safety factor for CF sheets (*BS EN 1992-1-1:2023*, Table J.1)

$$\gamma_f = 1.4$$

Reduction factor (*BS EN 1992-1-1:2023*, J.5.3 (2))

$$\eta_f = 0.7$$

$$f_{fud} = \frac{0.7 \times 3850}{1.4} = 1925\,\text{MPa}$$

Design ultimate strain (*BS EN 1992-1-1:2023*, Formula J.3)

$$\varepsilon_{fud} = \frac{f_{fud}}{E_f} = \frac{1925}{225\,000} = 0.00856 = 0.856\%$$

Width: $b_f = 450$ mm

The cross-sectional area of the fibres per metre of a single layer of CF sheet given by the supplier is 167 mm²/m.

Therefore

$$A_{f,1} = 0.167 \times 450 = 75.15\ \text{mm}^2$$

A maximum of 5 CFRP layers can be applied in accordance with *BS EN 1992-1-1:2023*, J.12.4 (1).

Effective depth of CFRP reinforcement (taken at the extreme concrete fibre)

$$d_f = 1225\,\text{mm}$$

Step 1 – Check bending moment capacity of existing section

Calculate the applied moment at ultimate limit state (ULS) after strengthening for partial factors of safety taken for CC2 – normal consequence class as defined in *BS EN 1990:2023*, Table 4.1.

Partial factor for permanent actions

$$\gamma_G = 1.35$$

Partial factor for variable actions

$$\gamma_Q = 1.5$$

Total factored load

$$1.35 \times 20 + 1.5 \times 50 = 102 \text{ kN/m}$$

Maximum ULS bending moment

$$M_{Ed} = \frac{102 \times 16^2}{8} = 3264 \text{kNm}$$

Maximum ULS shear force

$$V_{Ed} = \frac{102 \times 16}{2} = 816 \text{ kN}$$

Based on parabola-rectangle stress distribution in the compression zone given in *BS EN 1992-1-1:2023*, Figure 8.2c and using the iterative approach described in Annex A

$\varepsilon_{cu} = 0.0035 = 0.35\%$ (*BS EN 1992-1-1:2023*, Formula 8.4)

Neutral axis

$$x = 107.8 \text{ mm}$$

ULS moment capacity

$$M_{Rd,pre} = 2980 \text{kNm} < M_{Ed} = 3264 \text{kNm}$$

Therefore, the capacity of the existing section is not adequate for the applied moment at ULS and strengthening is required.

Step 2 – Calculate strains in steel and concrete at strengthening

The strengthening is applied under permanent loads – that is,

$$M_{Ed,\,str} = \frac{20 \times 16^2}{8} = 640 \text{ kNm}$$

Conservatively, the stress-strain analysis is performed assuming concrete is cracked in tension and using the parabola-rectangle stress distribution in the compression zone given in *BS EN 1992-1-1:2023*, Figure 8.2c, applying the iterative approach described in Annex A

$$x_{pre} = 293.9 \text{ mm}$$

Top surface strain (refer to Figure 13.2)

$$\varepsilon_{c,pre} = 0.00014 = 0.014\%$$

Soffit surface strain (refer to Figure 13.2)

$$\varepsilon_0 = 0.00044 = 0.045\%$$

Step 3 – Calculate required CFRP reinforcement

Based on initial calculations, 2 layers of CF sheets will be required. Therefore, in accordance with *BS EN 1992-1-1:2023*, J.5.2(4)

$$k_f = 1.0$$

Calculate the effective thickness (*BS EN 1992-1-1:2023*, Formula J.1)

$$t_f = \frac{n_f^{k_f} A_{f,1}}{b_f} = \frac{2^1 \times 75.15}{450} = 0.334 \text{ mm}$$

Cross-sectional area

$$A_f = 0.334 \times 450 = 150.3 \text{ mm}^2$$

Calculate the ULS moment capacity of the strengthened section assuming concrete fails first, based on parabola-rectangle stress distribution in the compression zone given in *BS EN 1992-1-1:2023*, Figure 8.2c, using the iterative approach described in Annex A.

$$x = 141.7 \text{ mm}$$

$$M_{Rd} = 3978 \text{kN}$$

Figure 13.2 T-beam strain distribution at strengthening – all dimensions are in mm

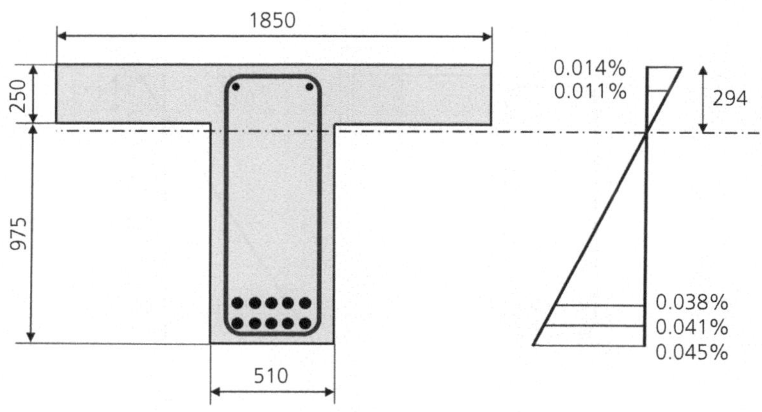

Check strain in CFRP

$$\varepsilon_f = 0.0262 = 2.62\% > \varepsilon_{fud} = 0.856\%$$

Therefore, the ULS moment capacity of the strengthened section will be limited by the CFRP ultimate strain.

$$x = 173.4 \text{ mm}$$

$$M_{Rd} = 3268\,\text{kNm} > M_{Ed} = 3264\,\text{kNm}$$

Top surface concrete strain

$$\varepsilon_c = 0.00148 = 0.148\% < \varepsilon_{cu} = 0.35\%$$

Maximum steel strain

$$\varepsilon_{st.1} = 0.00830 = 0.830\% < \frac{\varepsilon_{uk}}{\gamma_s} = \frac{5\%}{1.15} = 4.3\%$$

(elongation at maximum force assumed for Ductility Class B given in *BS EN 1992-1-1:2023*, Table C.2)

Therefore, 2 layers of CF sheets will be sufficient to provide the required moment capacity for the strengthened beam at ULS and the resulting strains are illustrated in Figure 13.3.

Step 4 – Determine anchorage requirements

As the surface tensile strength is not given, apply *BS EN 1992-1-1:2023*, Formula J.34.

$$f_{ctm,surf} = k_{c,surf} f_{ctm}$$

Figure 13.3 T-beam strain distribution at ULS for CFRP rupture – CF sheets strengthening – all dimensions are in mm

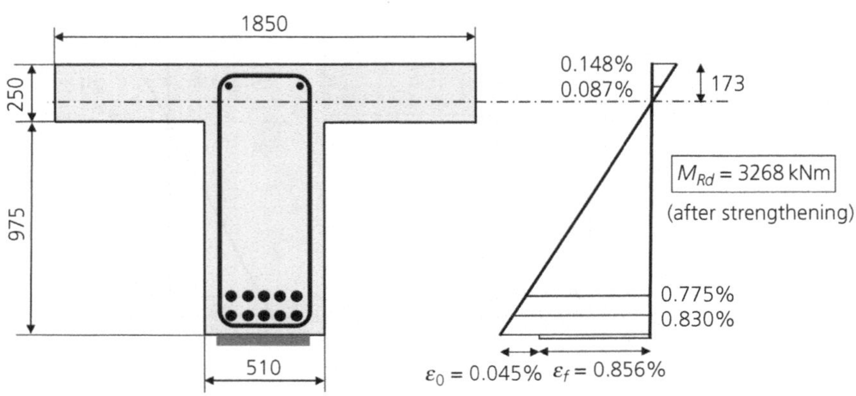

For bottom concreting position

$$k_{c,surf} = 0.6 + 0.3\left(\frac{f_{ck}}{60} - 0.2\right) = 0.6 + 0.3 \times \left(\frac{30}{60} - 0.2\right) = 0.69$$

$$f_{ctm,surf} = k_{c,surf} f_{ctm} = 0.69 \times 2.9 = 2.0 \text{ MPa}$$

Step 4.1 – End anchorage (*BS EN 1992-1-1:2023*, J.11.1.2.2)

The curtailment shall take account of a_l, calculated in accordance with *BS EN 1992-1-1:2023*, 12.3.2.

$$a_l = \frac{z\left(cot\theta - cot\alpha\right)}{2}$$

Where z is lever arm for the strengthened section (taken for d_{equiv} as defined in Equation 5.6).

$$d_{equiv} = \frac{d_f E_f A_f + \sum_{i=1}^{n} d_{st.i} E_s A_{st.i}}{E_f A_f + \sum_{i=1}^{n} E_s A_{st.i}}$$

$$= \frac{1225 \times 225\,000 \times 150.3 + (1143 + 1079) \times 200\,000 \times 4020}{225\,000 \times 150.3 + 200\,000 \times 8\,040} = 1113\,\text{mm}$$

It can be noted that the difference will be negligible if CFRP reinforcement was not taken into account – that is, the lever arm is calculated using

$$d_{st} = \frac{d_{st.1} + d_{st.2}}{2} = \frac{1143 + 1079}{2} = 1111\,\text{mm}$$

$cot\alpha = 0$ (vertical links)

Based on existing shear design

$cot\theta = 1.2$ $(\theta = 39.8°)$

Calculate the cracking moment in order to determine the minimum crack spacing and the cracked length of the beam, shown in Figure 13.4.

According to *BS EN 1992-1-1:2023*, the cracking moment may be calculated on the basis of the concrete tensile strength f_{ctm}. Using a simplified expression

$$M_{cr} = f_{ctm} \frac{I}{h_{max}}$$

Where

$I = 1.317 \times 10^{11}$ mm^4 – moment of inertia of concrete gross section

$h_{max} = 782.7$ mm – distance from centre of gravity to extreme concrete tension fibre.

Therefore

$$M_{cr} = 2.9 \times \frac{1.317 \times 10^{11}}{782.7} \times 10^{-6} = 488 \text{ kNm}$$

BS EN 1992-1-1:2023, Formula J.44 may be used to determine the minimum crack spacing.

$$s_{cr,min} = 1.5 \frac{M_{cr}}{0.85 h F_{fbsm}}$$

As the bars are ribbed, and given that they are cast in good conditions ($\kappa_{vb1} = 1.0$), the mean bond stress of the reinforcing steel may be determined by *BS EN 1992-1-1:2023*, Formula J.46.

$$f_{fbsm} = 0.43 f_{cm}^{\frac{2}{3}} = 0.43 \times 38^{\frac{2}{3}} = 4.86 \text{ MPa}$$

BS EN 1992-1-1:2023, Formula J.45 is used to determine the bond force per unit length of bars as follows

$$F_{fbsm} = \sum_{i=1}^{10} n_{s,i} \varnothing_i \pi f_{fbsm} = 10 \times 32 \times \pi \times 4.86 = 4886 \text{ N/mm}$$

Which gives the minimum crack spacing

$$s_{cr,min} = 1.5 \times \frac{488 \times 10^6}{0.85 \times 1225 \times 4886} = 144 \text{ mm}$$

Figure 13.4 shows the shifted moment diagram at a_l, indicating the point beyond which the CFRP strengthening is not required.

The existing capacity is sufficient to resist the applied ULS moment at distance 5.65 m from the support without taking into account the distance a_l. The calculated neutral axis at this point, under the applied bending moment after strengthening, is

At distance 5.65 m, $M_{Ed} = 2982.4 \text{ kNm}$, $x = 247.4$ mm

Figure 13.4 Applied and resistance bending moments including shift diagram

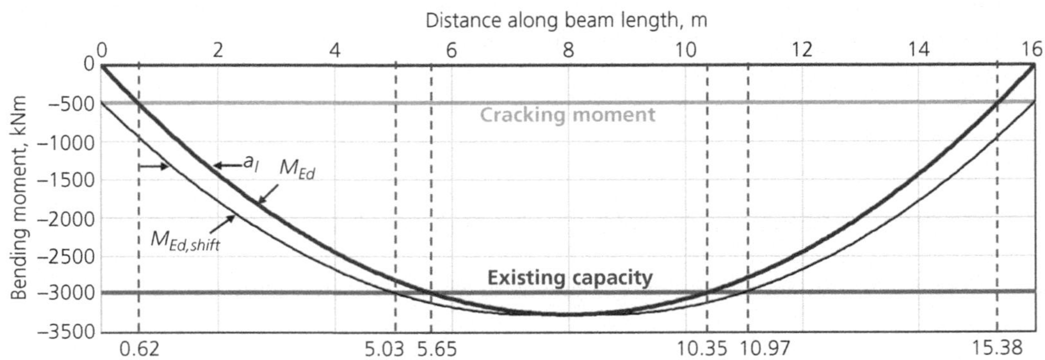

The corresponding distance from top of concrete to the concrete resultant force is 85.9 mm.

Therefore

$$z = 1113 - 85.9 = 1027.1 \text{ mm}$$

$$a_l = \frac{1027.1 \times 1.2}{2} = 616 \text{mm}$$

Similarly, the shift distance is calculated at each point along the beam length.

The anchorage resistance may be calculated using the simplified or refined method given in *BS EN 1992-1-1:2023*, J.11.1.1.2 and J.11.1.1.3, respectively.

To calculate the anchorage resistance using the refined method to *BS EN 1992-1-1:2023*, J.38

Assume $k_{sys,b1} = k_{sys,b2} = k_{sys,b3} = 1.0$

$$s_{f0k} = 0.2k_{sys,b2} = 0.2$$

$$\tau_{f1k} = 0.37k_{sys,b1}\sqrt{f_{cm}\, f_{ctm,surf}} = 0.37 \times \sqrt{38 \times 2.0} = 3.23 \text{ MPa}$$

$$l_{bf,max,k} = \frac{2}{k_{sys,b3}}\sqrt{\frac{E_f t_f s_{f0k}}{\tau_{f1k}}} = 2 \times \sqrt{\frac{225\,000 \times 0.334 \times 0.2}{3.23}} = 136.5 \text{mm}$$

$s_{cr,min} = 144 \text{ mm} > l_{bf,max,k}$, therefore the anchorage length will be limited by $l_{bf,max,k}$ and

$$f_{bfRd} = \frac{\sqrt{\eta_{cc} k_{tc} k_{tt}}}{\gamma_{BA}} f_{bfk,max}$$

$$f_{bfk,max} = \sqrt{\frac{E_f \tau_{f1k} s_{f0k}}{t_f}} = \sqrt{\frac{225\,000 \times 3.23 \times 0.2}{0.334}} = 659.3 \text{MPa}$$

$$f_{bfRd} = \frac{\sqrt{1.0 \times 0.85 \times 0.7}}{1.5} \times 659.3 = 339 \text{ MPa}$$

$$F_{bfRd} = f_{bfRd} A_f = 339 \times 150.3 = 50\,952 \text{ N} = 51 \text{ kN}$$

Since the anchorage region is within the cracked zone of the beam, an anchorage length of $2s_{cr,min}$ will be provided to ensure that the bond resistance can be generated between cracks.

As indicated in Figure 13.5, the resulting overall CFRP strengthening length is

$(11.8 - 4.2) + 2 \times (2 \times 0.144) = 8.18 \text{m}$ (at $\alpha_{fE} = 3.91 \text{m}$ distance from centre of support).

Figure 13.5 Force in CFRP and end anchorage resistance – CF sheets strengthening

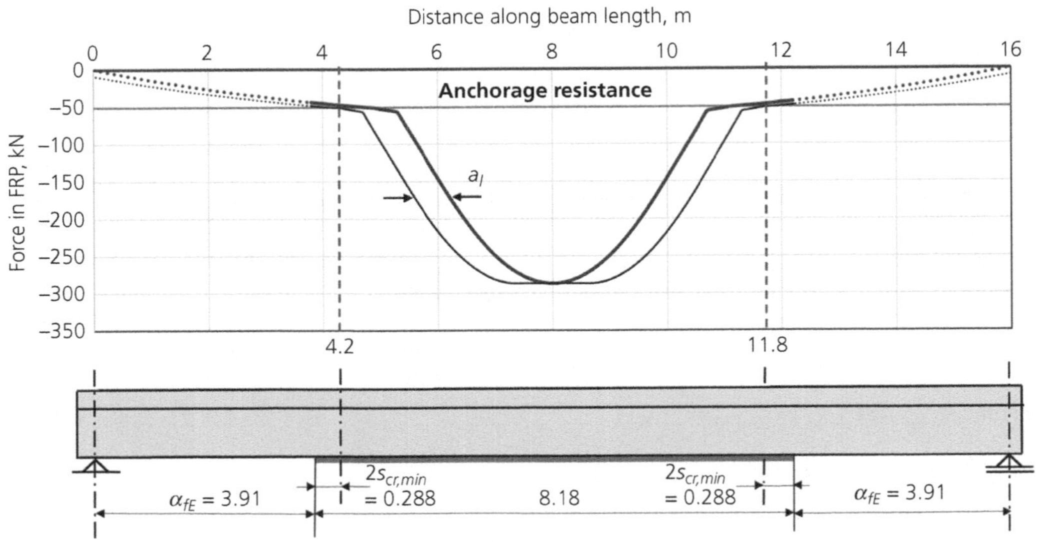

Step 4.2 – Intermediate crack debonding (*BS EN 1992-1-1:2023*, J.11.1.2.3)

In order to check the debonding between cracks, the tensile force in the CFRP has to be calculated at each predicted crack, assuming minimum crack spacing.

The analysis is performed using the iterative approach described in Annex A under the applied ULS moment. The force difference is calculated as per *BS EN 1992-1-1:2023*, Formula J.28.

$$\Delta f_{fEd} = \frac{F_{fEd,b} - F_{fEd,a}}{A_f}$$

Where a and b are the positions of two adjacent cracks.

The bond resistance should be calculated using *BS EN 1992-1-1:2023*, Formula J.47.

$$\Delta f_{fRd} = \frac{1}{\gamma_{BA}}\left[\sqrt{\eta_{cc}k_{tc}k_{tt}}\Delta f_{fk,B} + \Delta f_{fk,F} + \Delta f_{fk,C}\right]$$

Where

$\Delta f_{fk,B}$ is the term related to adhesive bond

$$\Delta f_{fk,B} = 0.84k_{sys,b1}\sqrt{f_{cm}f_{ctm,surf}}\frac{\sqrt{s_{cr,min}}}{t_f} = 0.84 \times 1.0 \times \sqrt{38 \times 2.0} \times \frac{\sqrt{144}}{0.334} = 263.2 \text{ MPa}$$

$\Delta f_{fk,F}$ is the term related to the beneficial effects of bond friction

$$\Delta f_{fk,F} = f_{cm}^{-0.9} \frac{s_{cr,min}^{\frac{4}{3}}}{t_f} = 38^{-0.9} \times \frac{144^{\frac{4}{3}}}{0.334} = 85.6 \text{ MPa}$$

$\Delta f_{fk,C}$ is the term related to clamping effect from curvature (only to be used for outward curve)

$$\Delta f_{fk,C} = \frac{\kappa_h}{h_f} \frac{s_{cr,min}^{\frac{1}{3}}}{t_f} = \frac{2\,000}{max(100,1225)} \times \frac{144^{\frac{1}{3}}}{0.334} = 25.6\,\text{MPa}$$

$$\Delta f_{fRd} = \frac{1}{1.5} \times \left[\sqrt{1.0 \times 0.85 \times 0.7} \times 263.2 + 85.6 + 25.6\right] = 135.3 + 57.0 + 17.1 = 209.4 \text{ MPa}$$

Conservatively, in this case the contribution of curvature may be ignored, as illustrated in Figure 13.6

$$\Delta f_{fRd} = 135.3 + 57.0 = 192.3 \text{ MPa}$$

Therefore, $\Delta f_{fEd} < \Delta f_{fRd}$ and the condition in *BS EN 1992-1-1:2023*, Formula J.43 is satisfied.

Table 13.1 presents the tabulated data at crack locations measured from midspan, including the first crack, corresponding to M_{cr} at distance 0.622 m from the support.

Figure 13.6 Stress in reinforcement (a) and force difference in CFRP (b) – CF sheet strengthening

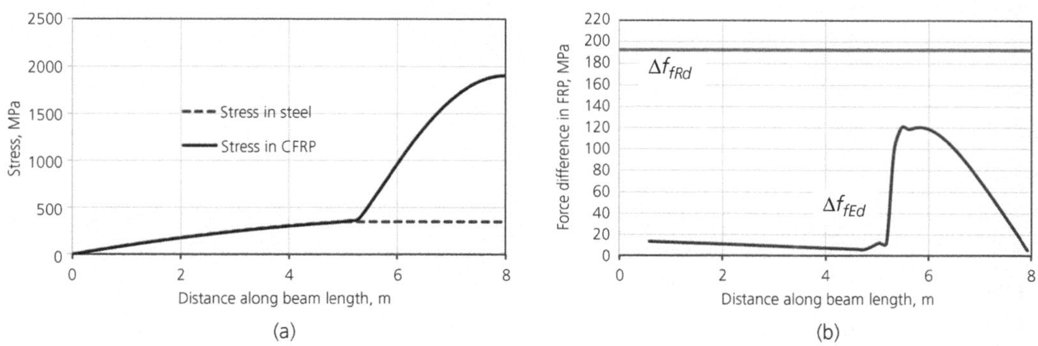

Step 4.3 – End cover separation check (*BS EN 1992-1-1:2023*, J.11.1.2.4)

Calculate the maximum shear force at the end of CFRP reinforcement at distance $a_{fE} = 3.91$ m from centre of support.

$$V_{Ed} = 417.2\,\text{kN}$$

$$V_{Rd,cfE} = \left(0.11 + 2.2 \frac{(100\rho_l)^{0.15}}{a_{fE}^{0.36}}\right)(100\rho_l f_{ck})^{\frac{1}{3}} b_w d_{st}$$

$$\rho_l = \frac{A_{sl}}{b_w d_{st}} = \frac{8040}{510 \times 1111} = 0.0142 \text{ (\textit{BS EN 1992-1-1:2023}, Formula 8.28)}$$

Table 13.1 Tabulated data – CF sheet strengthening (continued on next page)

Distance [m]	M_{Ed} [kNm]	V_{Ed} [kN]	x [mm]	a_l [mm]	F_{fEd} [kN]
8.000	3264	0	174.0	630	286.4
7.856	3263	15	174.1	630	285.6
7.712	3260	29	174.6	630	283.2
7.568	3254	44	175.3	630	279.2
7.424	3247	59	176.5	630	273.7
7.280	3238	73	177.9	629	266.6
7.136	3226	88	179.8	629	258.1
6.992	3212	103	182.1	629	248.0
6.848	3196	118	184.9	628	236.6
6.704	3178	132	188.4	628	223.7
6.560	3158	147	192.5	627	209.7
6.416	3136	162	197.5	626	194.4
6.272	3112	176	203.4	625	178.2
6.128	3085	191	210.6	623	161.0
5.984	3057	206	219.3	622	143.3
5.840	3026	220	229.9	620	125.2
5.696	2993	235	242.8	617	107.0
5.552	2958	250	258.7	614	89.1
5.408	2921	264	280.6	611	71.0
5.264	2882	279	305.0	607	56.2
5.120	2841	294	306.6	607	54.5
4.976	2798	308	308.4	607	52.6
4.832	2752	323	309.1	607	51.3
4.688	2705	338	308.8	607	50.4
4.544	2655	353	308.4	607	49.4
4.400	2603	367	308.1	607	48.5
4.256	2549	382	307.7	607	47.5
4.112	2493	397	307.3	607	46.4
3.968	2435	411	306.9	608	45.3
3.824	2375	426	306.5	608	
3.680	2312	441	306.1	608	
3.536	2248	455	305.7	608	
3.392	2181	470	305.3	608	
3.248	2112	485	304.9	608	
3.104	2041	499	304.4	608	

Table 13.1 Continued

Distance [m]	M_{Ed} [kNm]	V_{Ed} [kN]	x [mm]	a_l [mm]	F_{fEd} [kN]
2.960	1969	514	304.0	608	
2.816	1893	529	303.5	608	
2.672	1816	543	303.0	608	
2.528	1737	558	302.5	609	
2.384	1655	573	302.1	609	
2.240	1572	588	301.6	609	
2.096	1486	602	301.1	609	
1.952	1399	617	300.5	609	
1.808	1309	632	300.0	609	
1.664	1217	646	299.5	609	
1.520	1122	661	299.0	609	
1.376	1026	676	298.5	610	
1.232	928	690	297.9	610	
1.088	827	705	297.4	610	
0.944	725	720	296.8	610	
0.800	620	734	296.3	610	
0.656	513	749	295.7	610	
0.622	488	753	295.6	609	
0.000	0	816	293.8	611	

$$V_{Rd,cfE} = \left(0.11 + 2.2 \times \frac{(100 \times 0.0142)^{0.15}}{3910^{0.36}}\right) \times (100 \times 0.0142 \times 30)^{\frac{1}{3}} \times 510 \times 1111$$

$$V_{Rd,cfE} = 451\,159 \text{ N} = 451.2 \text{ kN}$$

$$V_{Ed} = 417.2\,\text{kN} < V_{Rd,cfE} = 451.2\,\text{kN}$$

Therefore, shear stirrup at the end of the longitudinal strengthening is not required.

Step 4.4 – Shear induced separation check (*BS EN 1992-1-1:2023*, J.11.1.2.5)

$$V_{Ed} = 417.2\,\text{kN}$$

Calculate the design shear stress at the end of CFRP reinforcement in accordance with *BS EN 1992-1-1:2023*, Formula 8.18.

$$\tau_{Ed} = \frac{V_{Ed}}{b_w z} = \frac{V_{Ed}}{b_w 0.9 d_{st}} = \frac{417.2 \times 10^3}{510 \times 0.9 \times 1111} = 0.82 \text{ MPa}$$

Check if the limit in *BS EN 1992-1-1:2023*, Formula J.54 is exceeded.

$$\tau_{Ed} = 0.82 \text{ MPa} < 0.33 f_{ck}^{\frac{2}{3}} = 0.33 \times 30^{\frac{2}{3}} = 3.2 \text{ MPa} - \text{check is ok.}$$

The shear resistance in this case can be taken as the upper limit of *BS EN 1992-1-1:2023*, Formula 8.62.

$$\tau_{Rd} = \nu f_{cd} \frac{cot\theta}{1 + cot^2\theta} = 0.5 \times 17 \times \frac{1.2}{1 + 1.2^2} = 4.18 \text{ MPa}$$

Conservatively, take

$$\sigma_{swd} = f_{ywd} = 348 \text{ MPa}$$

For ribbed steel bars (*BS EN 1992-1-1:2023*, Formula J.53)

$$\frac{\tau_{Ed}\sigma_{swd}}{\tau_{Rd}} = \frac{0.82 \times 348}{4.18} = 68 \text{ MPa} < 75 \text{ MPa}$$

Therefore, additional EBR stirrups are not required.

Example Flexural Design B-2 (CFRP strips)

Calculate the ULS bending moment capacity of the T-beam described in Example Flexural Design B-1 for EBR strengthening using CFRP strips.

CFRP strips properties

Characteristic short-term tensile strength (5% fractile): $f_{fuk} = 2900 \text{ MPa}$

Mean modulus of elasticity: $E_f = 170\,000 \text{ MPa}$

Characteristic ultimate strain at rupture: $\varepsilon_{fuk} = \frac{f_{fuk}}{E_f} = \frac{2900}{170\,000} = 0.0171 = 1.71\%$

Design tensile strength (*BS EN 1992-1-1:2023*, Formula J.2): $f_{fud} = \frac{\eta_f f_{fuk}}{\gamma_f}$

Partial safety factor (*BS EN 1992-1-1:2023*, Table J.1): $\gamma_f = 1.3$

Reduction factor (*BS EN 1992-1-1:2023*, J.5.3 (2)): $\eta_f = 0.7$

$$f_{fud} = 0.7 \times \frac{2900}{1.3} = 1562 \text{ MPa}$$

Design ultimate strain (*BS EN 1992-1-1:2023*, Formula J.3)

$$\varepsilon_{fud} = \frac{f_{fud}}{E_f} = \frac{1562}{170\,000} = 0.0919 = 0.919\%$$

The strip width is 150 mm.

The thickness of a single strip layer is 1.2 mm.

Assume 2 CFRP strips installed at 250 mm centre-to-centre spacing. Therefore

$$A_f = 2\times1.2\times150 = 360\,\text{mm}^2$$

A maximum of 2 layers of CFRP strips can be applied in accordance with *BS EN 1992-1-1:2023*, J.12.4 (2) with a total thickness not exceeding 3 mm. The minimum edge distance should be at least equivalent to the nominal concrete cover to internal reinforcement in accordance with *BS EN 1992-1-1:2023*, J.12.1(2). The maximum spacing according to *BS EN 1992-1-1:2023*, J.12.1(2) in this case is 400 mm.

Effective depth of CFRP reinforcement (taken at the extreme concrete fibre)

$$d_f = 1225\,\text{mm}$$

Step 1 – Check bending moment capacity of existing section

As per Example Flexural Design B-1.

Step 2 – Calculate strains in steel and concrete at strengthening

As per Example Flexural Design B-1.

Step 3 – Calculate required CFRP reinforcement

Firstly, calculate the ULS capacity assuming concrete reaches its ultimate strain of 0.0035. Assuming parabola-rectangle stress distribution in the compression zone (*BS EN 1992-1-1:2023*, Figure. 8.2c), use the iterative approach described in Annex A.

$$x = 161.2\text{ mm}$$

$$M_{Rd} = 4524.2\,\text{kNm}$$

Check strain in CFRP

$$\varepsilon_f = 0.02265 = 2.27\% > \varepsilon_{fud} = 0.919\%$$

Therefore, the ULS moment capacity of the strengthened section will be limited by the CFRP ultimate strain.

$$x = 177.5\text{ mm}$$

$$M_{Rd} = 3\,578\text{ Nm} > M_{Ed} = 3\,264\text{ kN}$$

Top surface concrete strain

$$\varepsilon_c = 0.00163 = 0.163\% < \varepsilon_{cu} = 0.35\%$$

Maximum steel strain

$$\varepsilon_{st.1} = 0.00888 = 0.888\% < \frac{\varepsilon_{uk}}{\gamma_s} = \frac{5\%}{1.15} = 4.3\%$$

(elongation at maximum force assumed for Ductility Class B given in *BS EN 1992-1-1:2023*, Table C.2)

Therefore, a single layer of 2 CFRP strips will be sufficient to provide the required moment capacity of the strengthened beam at ULS and the resulting strains are illustrated in Figure 13.7. (Note

Figure 13.7 Strain distribution at ULS for CFRP rupture – CFRP strips strengthening – all dimensions are in mm

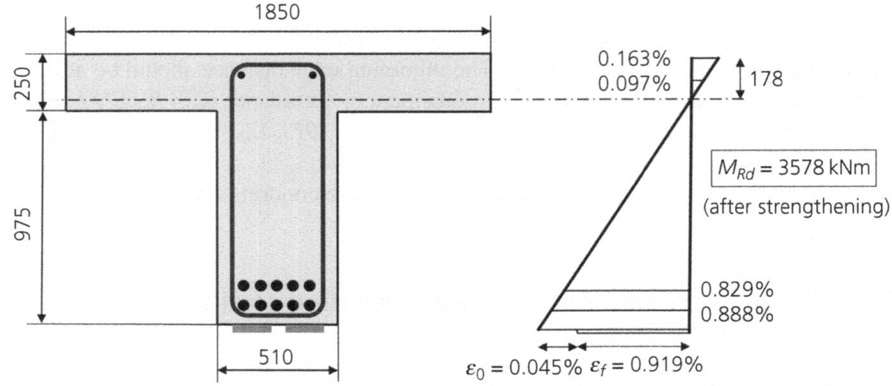

that the minimum required strengthening capacity in this example is governed by the intermediate crack debonding in Step 4.2.)

Step 4 – Determine anchorage requirements

Calculate surface tensile strength as per Example Flexural Design B-1.

Step 4.1 – End anchorage (*BS EN 1992-1-1:2023*, J.11.1.2.2)

Moment diagram, cracking moment and minimum crack spacing are as per Example Flexural Design B-1.

Calculate anchorage resistance to *BS EN 1992-1-1:2023*, J.38 (refined method).

All assumptions are as per Example Flexural Design B-1.

$$l_{bf,max,k} = \frac{2}{k_{sys,b3}}\sqrt{\frac{E_f t_f s_{f0k}}{\tau_{f1k}}} = 2\times\sqrt{\frac{170\,000\times1.2\times0.2}{3.23}} = 224.9\text{ mm}$$

$$f_{bfk,max} = \sqrt{\frac{E_f \tau_{f1k} s_{f0k}}{t_f}} = \sqrt{\frac{170\,000\times3.23\times0.2}{1.2}} = 302.3\text{ MPa}$$

$$s_{cr,min} = 144\text{ mm} < l_{bf,max,k}$$

Therefore, the anchorage length will be limited by $s_{cr,min}$.

For $l_{bf} = s_{cr,min} = 144\text{ mm} < l_{bf,max,k}$

$$f_{bfRd} = \frac{\sqrt{\eta_{cc}k_{tc}k_{tt}}}{\gamma_{BA}} f_{bfk,max}\frac{l_{bf}}{l_{bf,max,k}}\left(2-\frac{l_{bf}}{l_{bf,max,k}}\right)$$

$$f_{bfRd} = \frac{\sqrt{1.0\times0.85\times0.7}}{1.5}\times302.3\times\frac{144}{224.9}\times\left(2-\frac{144}{224.9}\right) = 135.4\text{ MPa}$$

$$F_{bfRd} = f_{bfRd}A_f = 135.4\times360 = 48\,730\text{ N} = 48.7\text{ kN}$$

In this case a_l is calculated using

$$d_{equiv} = \frac{1225 \times 170\,000 \times 360 + (1143 + 1079) \times 200\,000 \times 4020}{170\,000 \times 360 + 200\,000 \times 8040} = 1115 \text{ mm}$$

Since the anchorage region is within the cracked zone of the beam, an anchorage length of $2s_{cr,min}$ will be provided to ensure that the bond resistance can be generated between cracks.

As indicated in Figure 13.8, the resulting overall CFRP strengthening length is $(14.51 - 1.49) + (2 \times 0.144) = 13.6\,\text{m}$ (at $\alpha_{fE} = 1.36\,\text{m}$ (at $\alpha_{fE} = 1.2$ m distance from centre of support).

Step 4.2 – Intermediate crack debonding (*BS EN 1992-1-1:2023*, J.11.1.2.3)
Calculate bond resistance using *BS EN 1992-1-1:2023*, Formula J.47 as per Example Flexural Design B-1 for CFRP thickness of 1.2 mm.

$$\Delta f_{fk,B} = 0.84 k_{sys,b1} \sqrt{f_{cm} f_{ctm,surf}} \frac{\sqrt{s_{cr,min}}}{t_f} = 0.84 \times 1 \times \sqrt{38 \times 2.0} \times \frac{\sqrt{144}}{1.2} = 73.2 \text{ MPa}$$

$$\Delta f_{fk,F} = f_{cm}^{-0.9} \times \frac{s_{cr,min}^{\frac{4}{3}}}{t_f} = 38^{-0.9} \times \frac{144^{\frac{4}{3}}}{1.2} = 23.8\,\text{MPa}$$

$$\Delta f_{fk,C} = \frac{\kappa_h}{h_f} \times \frac{s_{cr,min}^{\frac{1}{3}}}{t_f} = \frac{2000}{1220} \times \frac{144^{\frac{1}{3}}}{1.2} = 7.1\,\text{MPa}$$

$$\Delta f_{fRd} = \frac{1}{1.5} \times \left[\sqrt{1.0 \times 0.85 \times 0.7} \times 73.2 + 23.8 + 7.1\right] = 37.7 + 15.9 + 4.8 = 58.4 \text{ MP}$$

Figure 13.8 Force in CFRP and end anchorage resistance – CFRP strips strengthening

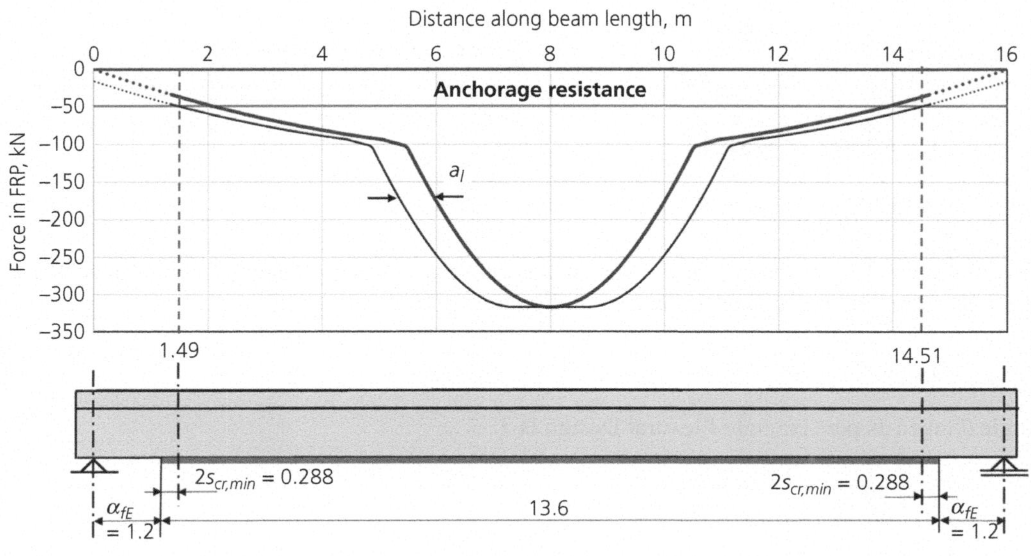

Maximum $\Delta f_{fEd} = 58.2$ MPa $< \Delta f_{fRd} = 58.4$ MPa, only when taking into account the curvature term (refer to Figure 13.9).

Therefore, the condition in *BS EN 1992-1-1:2023*, Formula J.43 is satisfied.

Table 13.2 presents the tabulated data at crack locations measured from midspan, including the first crack, corresponding to M_{cr} at distance 0.622 m from the support.

Figure 13.9 Force in CFRP and steel (a) and force difference in CFRP (b) – CFRP strips

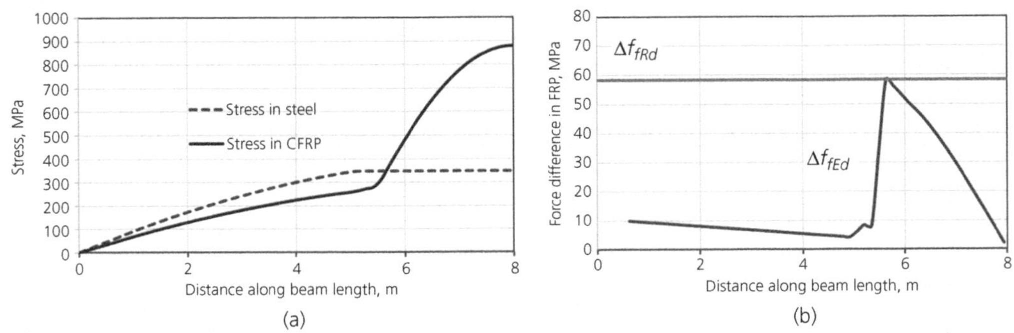

Step 4.3 – End cover separation check

Calculate the maximum shear force at the end of CFRP reinforcement at distance $\alpha_{fE} = 1.2$ m from centre of support.

$$V_{Ed} = 693.6\,\text{kN}$$

As per Example Flexural Design B-1

$$V_{Rd,cfE} = \left(0.11 + 2.2 \times \frac{(100 \times 0.0142)^{0.15}}{1200^{0.36}}\right) \times (100 \times 0.0142 \times 30)^{\frac{1}{3}} \times 510 \times 1111$$

$$V_{Rd,cfE} = 574921\,\text{N} = 575\,\text{kN}$$

$$V_{Ed} = 693.6\,\text{kN} > V_{Rd,cfE} = 575\,\text{kN}$$

Therefore, shear strengthening at the end of the longitudinal strengthening shall be provided to resist the force calculated in accordance with Equation 3.16.

$$F_{st} = F_{bfRd} tan\theta$$

Angle θ taken as per Example Flexural Design B-1.

$$F_{st} = 48.7 \times tan 39.8° = 40.6\ \text{kN}$$

Table 13.2 Tabulated data – CFRP strips strengthening (continued on next page)

Distance [m]	M_{Ed} [kNm]	V_{Ed} [kN]	x [mm]	a_l [mm]	F_{fEd} [kN]
8.000	3264	0	209.5	625	316.2
7.857	3263	15	209.7	625	315.4
7.713	3260	29	210.2	624	313.1
7.570	3255	44	211.0	624	309.2
7.427	3247	58	212.2	624	303.8
7.284	3238	73	213.8	624	296.8
7.140	3226	88	215.8	623	288.3
6.997	3213	102	218.2	623	278.4
6.854	3197	117	221.2	622	267.1
6.710	3179	132	224.8	622	254.4
6.567	3159	146	229.0	621	240.4
6.424	3137	161	234.0	620	225.2
6.280	3113	175	239.9	619	208.8
6.137	3087	190	246.9	617	191.5
5.994	3059	205	255.2	616	173.3
5.851	3028	219	265.7	614	154.1
5.707	2996	234	279.0	612	133.9
5.564	2961	248	296.3	610	112.9
5.421	2925	263	307.8	608	100.6
5.277	2886	278	309.3	608	97.7
5.134	2845	292	310.9	608	94.7
4.991	2802	307	311.5	608	92.5
4.847	2757	322	311.2	608	91.0
4.704	2710	336	310.8	608	89.4
4.561	2661	351	310.5	608	87.7
4.418	2609	365	310.1	608	86.0
4.274	2556	380	309.8	608	84.2
4.131	2501	395	309.4	608	82.3
3.988	2443	409	309.0	609	80.4
3.844	2383	424	308.6	609	78.4
3.701	2321	438	308.2	609	76.3
3.558	2258	453	307.8	609	74.1
3.414	2192	468	307.4	609	71.9
3.271	2124	482	306.9	609	69.6

Table 13.2 Continued

Distance [m]	M_{Ed} [kNm]	V_{Ed} [kN]	x [mm]	a_l [mm]	F_{fEd} [kN]
3.128	2053	497	306.5	609	67.3
2.985	1981	512	306.0	609	64.9
2.841	1907	526	305.5	609	62.4
2.698	1830	541	305.1	609	59.8
2.555	1752	555	304.6	610	57.2
2.411	1671	570	304.1	610	54.5
2.268	1588	585	303.6	610	51.7
2.125	1504	599	303.1	610	48.9
1.981	1417	614	302.6	610	46.0
1.838	1328	629	302.1	610	43.0
1.695	1236	643	301.6	610	40.0
1.552	1143	658	301.0	610	36.9
1.408	1048	672	300.5	610	33.7
1.265	951	687	300.0	611	30.5
1.122	851	702	299.4	611	
0.978	749	716	298.9	611	
0.835	646	731	298.3	611	
0.692	540	745	297.8	611	
0.622	488	753	297.6	609	
0.000	0	816	296.3	612	

Calculate f_{fwd} for open continuous CFRP systems as per *BS EN 1992-1-1:2023*, Formula J.23.

Assume CF sheets as for flexural strengthening in Example Flexural Design B-1, assuming $\alpha_f = 90°$.

For a metre width

$A_{f,1} = 167\ \text{mm}^2$

Use 4 layers with effective thickness

$t_f = \dfrac{n_f^{k_f} A_{f,1}}{2} = \dfrac{4^{0.85} \times 167}{1000} = 0.543\,\text{mm}$ (calculated as per *BS EN 1992-1-1:2023*, Formula J.1 with

$E_f = 225\,000$ MPa).

Therefore

$$l_{bf,max,k} = \frac{2}{k_{sys,b3}}\sqrt{\frac{E_f t_f s_{f0k}}{\tau_{f1k}}} = 2\times\sqrt{\frac{225\,000\times0.543\times0.2}{3.23}} = 173.9 \text{ mm}$$

$$f_{bfRd} = \frac{\sqrt{\eta_{cc}k_{tc}k_{tt}}}{\gamma_{BA}} f_{bfk,max} = \frac{\sqrt{\eta_{cc}k_{tc}k_{tt}}}{\gamma_{BA}}\sqrt{\frac{E_f \tau_{f1k} s_{f0k}}{t_f}}$$

$$= \frac{\sqrt{1.0\times0.85\times0.7}}{1.5}\times\sqrt{\frac{225\,000\times3.23\times0.2}{0.543}} = 266\,\text{MPa}$$

$h_f = 975$ mm (height of shear strengthening as per definition in *BS EN 1992-1-1:2023*, Figure J.4)

$$f_{fwd} = \begin{cases} \dfrac{2}{3}\dfrac{\frac{h_f}{sin\alpha_f}}{l_{bf,max,k}} f_{bfRd} & \text{for } \dfrac{h_f}{sin\alpha_f} < l_{bf,max,k} \\ \left(1-\dfrac{1}{3}\dfrac{l_{bf,max,k}}{\frac{h_f}{sin\alpha_f}}\right) f_{bfRd} & \text{for } \dfrac{h_f}{sin\alpha_f} \geq l_{bf,max,k} \end{cases}$$

$$f_{fwd} = \left(1-\frac{1}{3}\frac{l_{bf,max,k}}{h_f}\right) f_{bfRd} = \left(1-\frac{1}{3}\times\frac{173.9}{975}\right)\times 266 = 250 \text{ MPa}$$

The required area of end stirrup is

$$A_f = \frac{40.6}{250}\times10^3 = 162.2 \text{ mm}^2$$

The required width per leg is

$$b_f = \frac{162.2}{2\times0.543} = 149.5 \text{ mm}$$

2 legs of 4 layers of continuous CF sheets with 150 mm width will be adequate, as indicated in Figure 13.10.

Figure 13.10 End shear stirrup to prevent end cover separation – CFRP strips strengthening – all dimensions are in mm

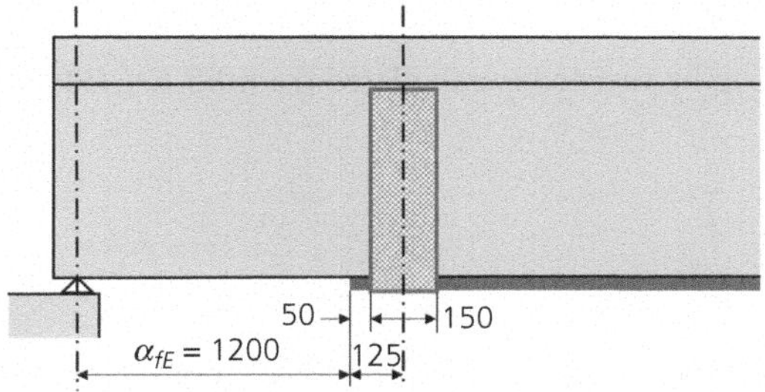

Step 4.4 – Shear induced separation check (*BS EN 1992-1-1:2023*, J.11.1.2.5)

$$V_{Ed} = 693.6\,\text{kN}$$

$$M_{Ed} = 905.8\,\text{kNm}$$

Calculate the design shear stress at the end of CFRP reinforcement in accordance with *BS EN 1992-1-1:2023*, Formula 8.18.

$$\tau_{Ed} = \frac{V_{Ed}}{b_w z} = \frac{V_{Ed}}{b_w 0.9 d_{st}} = \frac{693.6\times10^3}{510\times0.9\times1111} = 1.36\ \text{MPa}$$

Check if the limit in *BS EN 1992-1-1:2023*, Formula J.54 is exceeded.

$$\tau_{Ed} = 1.36\ \text{MPa} < 0.33 f_{ck}^{\frac{2}{3}} = 0.33\times30^{\frac{2}{3}} = 3.2\ \text{MPa} - \text{check is ok.}$$

$$\tau_{RD} = 4.18\,\text{MPa} - \text{as per Example Flexural Design B-1}$$

Conservatively, take

$$\sigma_{swd} = f_{ywd} = 348\,\text{MPa}$$

For ribbed steel bars (*BS EN 1992-1-1:2023*, Formula J.53)

$$\frac{\tau_{Ed}\sigma_{swd}}{\tau_{Rd}} = \frac{1.36\times348}{4.18} = 113\,\text{MPa} > 75\,\text{MPa}$$

As the check is undertaken outside of the region close to the support, the stress in the steel stirrups may be calculated by equally distributing the applied shear force into each leg of each stirrup that crosses the shear crack at the prescribed angle.

$$\sigma_{swd} = \frac{V_{Ed}}{\left(\frac{A_{sw}}{s}\right) z\left(cot\theta + cot\alpha\right) sin\alpha} = \frac{693.6\times10^3}{\left(\frac{402}{200}\right)\times0.9\times1111\times\left(1.2+0\right)\times1} = 287.5\ \text{MPa}$$

$$\frac{\tau_{Ed}\sigma_{swd}}{\tau_{Rd}} = \frac{1.36\times287.5}{4.18} = 93.5\,\text{MPa} > 75\,\text{MPa}$$

Therefore, additional stirrups are required. They need to extend over the whole length exceeding the stress limit. Using the more conservative approach, the applied shear force needs to be reduced to

$V_{Ed} = \frac{75}{113}\times693.6 = 460.4$ kN, which corresponds to 3.5 m from the end support.

The additional shear force to be resisted is calculated from *BS EN 1992-1-1:2023*, Formula J.55.

$$\tau_{Rd,sy} = \rho_w f_{ywd} cot\theta$$

Where

$$\rho_w = \frac{A_{sw}}{b_w s} = \frac{2\times\pi\times16^2}{510\times200} = 0.0039\ \ (\textit{BS EN 1992-1-1:2023}, \text{Formula 8.43})$$

Therefore

$$\tau_{Rd,sy} = 0.0039 \times 348 \times 1.2 = 1.65 \text{ MPa} > \tau_{Ed} = 1.36 \text{ MPa}$$

Then,

$$\tau_{Ed,f} = \frac{E_f A_f}{E_f A_f + E_s A_{st}} \tau_{Ed} = \frac{170\,000 \times 360}{170\,000 \times 360 + 200\,000 \times 8040} \times 1.36 = 0.05 \text{MPa}$$

Consider shear strengthening with open continuous CFRP systems as for the end stirrup.

Use the same CF sheets as for the end shear stirrup, assuming $\alpha_f = 90°$.

However, in this case use a single layer with thickness $t_f =$ 0.167 mm.

Therefore

$$l_{bf,max,k} = 2 \times \sqrt{\frac{225\,000 \times 0.167 \times 0.2}{3.23}} = 96.5 \text{mm}$$

$$f_{bfRd} = \frac{\sqrt{1.0 \times 0.85 \times 0.7}}{1.5} \times \sqrt{\frac{225\,000 \times 3.23 \times 0.2}{0.167}} = 480 \text{MPa}$$

$$f_{fwd} = \left(1 - \frac{1}{3} \times \frac{96.5}{975}\right) \times 480 = 464 \text{ MPa}$$

$$\tau_{Rd,f} = 2t_f \frac{f_{fwd}}{b_w} = 2 \times 0.167 \times \frac{464}{510} = 0.30 \text{ MPa} > \tau_{Ed,f} = 0.05 \text{MPa}$$

Therefore, the proposed CFRP shear strengthening is adequate.

As the shear strengthening overlaps with the end strap required to prevent end cover separation, it is necessary to superimpose both effects and re-evaluate the effective thickness of the CFRP system for the increased number of layers.

For 5 layers, the effective thickness is

$$t_f = \frac{n_f^{k_f} A_{f,1}}{b_f} = \frac{5^{0.85} \times 167}{1000} = 0.656 \text{ mm}$$

Capacity per 2 legs of 150 mm wide end strap for 5 layers of CF sheets

$$2f_{fwd} t_f b_f = 2 \times 250 \times 0.656 \times 150 = 49\,200 \text{ N} = 49.2 \text{ kN}$$

Additional force in the end strap due to $\tau_{Ed,f} = 0.05$ MPa

$$2\tau_{Ed,f} t_f b_f = 2 \times 0.05 \times 0.656 \times 150 = 10 \text{ N} = 0.01 \text{ kN}$$

Therefore, no significant change of the total force in the end strap equal to

$$40.6 + 0.01 = 40.61 \text{ kN}$$

$$40.61 \text{ kN} < 49.2 \text{ kN} - \text{check is ok}$$

In comparison with Example Flexural Design B-1, the use of CFRP strips is less efficient than CF sheets for this particular design. This is due to the lower bond resistance of the CFRP strip system at intermediate cracks, which also results in higher applied shear forces in the anchorage zones and requirement for additional shear strengthening.

Example Flexural Design B-3 (NSM CFRP)

Calculate the ULS bending moment capacity of the T-beam described in Example Flexural Design B-1 for NSM strengthening using CFRP strips.

CFRP strips properties

Characteristic short-term tensile strength (5 % fractile): $f_{fuk} = 2900$ MPa

Mean modulus of elasticity: $E_f = 170\,000$ MPa

Characteristic ultimate strain at rupture: $\varepsilon_{fuk} = \dfrac{f_{fuk}}{E_f} = \dfrac{2900}{170\,000} = 0.0171 = 1.71\%$

Design tensile strength (*BS EN 1992-1-1:2023*, Formula J.2): $f_{fud} = \dfrac{\eta_f f_{fuk}}{\gamma_f}$

Partial safety factor (*BS EN 1992-1-1:2023*, Table J.1): $\gamma_f = 1.3$

Reduction factor (*BS EN 1992-1-1:2023*, J.5.3 (2)): $\eta_f = 0.7$

$$f_{fud} = \frac{0.7 \times 2900}{1.3} = 1562 \text{ MPa}$$

Design ultimate strain (*BS EN 1992-1-1:2023*, Formula J.3)

$$\varepsilon_{fud} = \frac{f_{fud}}{E_f} = \frac{1562}{170000} = 0.00919 = 0.919\%$$

The thickness of NSM strip is: $t_f = 2.5$ mm.

The width of NSM strip: $b_f = 20$ mm.

Therefore, the area of NSM strip is 50 mm^2.

No more than a single NSM strip or bar should be bonded into one slot in accordance with *BS EN 1992-1-1:2023*, J.12.4 (3).

In accordance with *BS EN 1992-1-1:2023*, Table J.2 for NSM CFRP strips

Slot width: $(t_f + 2) \leq b_{slot} \leq (t_f + 4)$ or 4.5 mm $\leq b_{slot} \leq 6.5$ mm

Slot depth: $b_f \leq t_{slot} \leq (b_f + 2)$ or $20\text{ mm} \leq t \leq 22\text{ mm}$

Assume: $b_{slot} = 6$ mm and $t_{slot} = 22$ mm

Edge distance (from centre of slot to edge): $a_r \geq 4\ b_f = 80$ mm

Centre-to-centre spacing: $max\left\{b_f; D_{upper}\right\} \leq s_f \leq min\left\{0.2l_{0b}; 3h\right\}$

Where

l_{0b} is the distance between points of zero moment in mm

D_{upper} is the upper sieve size of the coarsest fraction of aggregates used in concrete

$$max\{20;40\} \leq s_f \leq min\{0.2 \times 16000; 3 \times 1220\}$$

$$max\{20;40\} \leq s_f \leq min\{3200;3660\}$$

$$40\,\text{mm} \leq s_f \leq 3200\,\text{mm}$$

Try 4 NSM CFRP strips at 100 mm centre-to-centre spacing.

$$A_f = 4 \times 50 = 200\ \text{mm}^2$$

Effective depth of CFRP reinforcement: $d_f = 1225 - 22 + 10 = 1213$ mm

Adhesive properties:

Characteristic tensile strength of the adhesive: $f_{Atk} = 30$ MPa

Characteristic compressive strength of the adhesive: $f_{Ack} = 90$ MPa

$$\alpha_{bA} = 0.5$$

$$\alpha_{bc} = \sqrt{\eta_{cc} k_{tc} k_{tt}} = 1.0 \times 0.85 \times 0.7 = 0.771$$

Bond strength of adhesive with NSM CFRP reinforcement strips (*BS EN 1992-1-1:2023*, Formula J.60)

$$\tau_{bAk} = 0.6\sqrt{\left(2f_{Atk} - 2\sqrt{f_{Atk}^2 + f_{Ack} f_{Atk}} + f_{Ack}\right) f_{Atk}}$$

$$\tau_{bAk} = 0.6\sqrt{\left(2 \times 30 - 2 \times \sqrt{30^2 + 30 \times 90} + 90\right) \times 25} = 18\ \text{MPa}$$

Bond strength of concrete with NSM CFRP reinforcement strips (*BS EN 1992-1-1:2023*, Formula J.61)

$$\tau_{bck} = 4.5\sqrt{f_{cm}} = 4.5\sqrt{38} = 27.7\ \text{MPa}$$

$$\tau_{bAk}\alpha_{bA} = 9\ \text{MPa} < \tau_{bck}\alpha_{bc} = 21.4\ \text{MPa}$$

Design value of the shear stress resistance of adhesive (*BS EN 1992-1-1:2023*, Formula J.51)

$$\tau_{bAd} = \frac{1}{\gamma_{BA}} \tau_{bAk} \alpha_{bA} = \frac{9}{1.5} = 6 \text{ MPa}$$

Step 1 – Check bending moment capacity of existing section

As per Example Flexural Design B-1.

Step 2 – Calculate strains in steel and concrete at strengthening

As per Example Flexural Design B-1.

Step 3 – Calculate required CFRP reinforcement

Firstly, calculate the ULS capacity assuming concrete reaches its ultimate strain of 0.0035. Assuming parabola-rectangle stress distribution in the compression zone (*BS EN 1992-1-1:2023*, Figure 8.2c)

$$x = 141.6 \text{ mm}$$

$$M_{Rd} = 3963 \text{ kNm}$$

Check strain in CFRP

$$\varepsilon_f = 0.02605 = 2.61\ \% > \varepsilon_{fud} = 0.919\ \%$$

Therefore, the ULS moment capacity of the strengthened section will be limited by the CFRP ultimate strain. Assuming parabola-rectangle stress distribution in the compression zone (*BS EN 1992-1-1:2023*, Figure 8.2c)

$$x = 169.1 \text{ mm}$$

$$M_{Rd} = 3295 \text{ kNm} > M_{Ed} = 3264 \text{ kN}$$

Top surface concrete strain

$$\varepsilon_c = 0.00156 = 0.156\ \% < \varepsilon_{cu} = 0.35\ \%$$

Maximum steel strain

$$\varepsilon_{st.1} = 0.00888 = 0.898\% < \frac{\varepsilon_{uk}}{\gamma_s} = \frac{5\%}{1.15} = 4.3\%$$

(elongation at maximum force assumed for Ductility Class B given in *BS EN 1992-1-1:2023*, Table C.2)

Therefore, four NSM CFRP strips will be sufficient to provide the required moment capacity of the strengthened beam at ULS and the resulting strains are illustrated in Figure 13.11.

Figure 13.11 Strain distribution at ULS for CFRP rupture – NSM strengthening – all dimensions are in mm

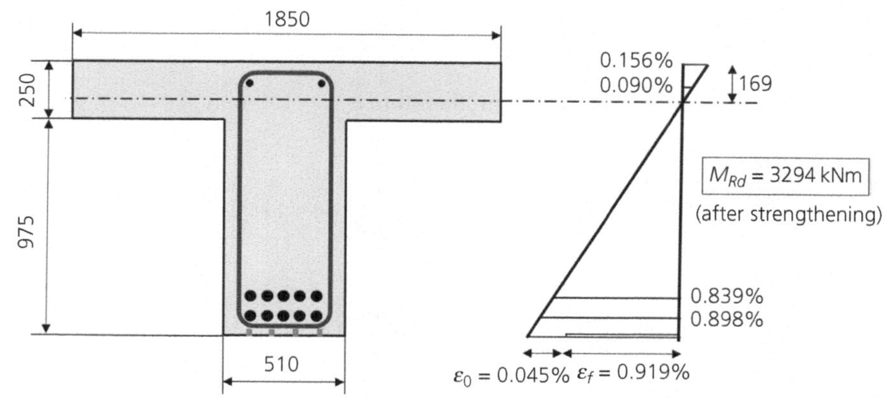

Step 4 – Determine anchorage requirements

Step 4.1 – End anchorage (*BS EN 1992-1-1:2023*, J.11.1.3)

The NSM CFRP strengthening will be anchored in the zone where steel has not yielded yet.

Figure 13.12 indicates that the end anchorage may be in the cracked zone, in which case the end anchorage length is limited by the minimum crack spacing, as calculated in Example Flexural Design B – 1 – that is,

$$l_{bf} = 144\,\text{mm} > 115\,\text{mm}$$

Anchorage force resistance in each NSM strip

$$F_{bfRd,i} = 0.95 b_f \tau_{bAd} a_r^{\frac{1}{4}} \left(26.2 + 0.065 tanh\left(\frac{a_r}{70}\right)(l_{bf} - 115)\right)$$

Figure 13.12 Applied and resistance bending moments including shift diagram – NSM strengthening

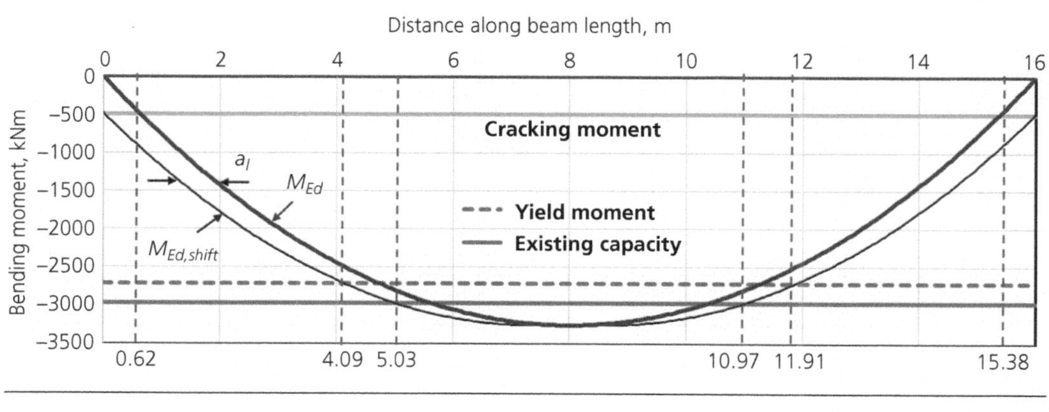

Calculate the distance from the centre of each NSM strip assuming they are spaced equally and have equal edge distances.

Outer strips: $a_{r,1} = \frac{(510 - 3 \times 100)}{2} = 105$ mm

Inner strips: $a_{r,2} = 105 + 100 = 205 > 150$ mm, therefore $a_{r,2} = 150$ mm

Thus,

$$F_{bfRd,1} = 0.95 \times 20 \times 6 \times 105^{\frac{1}{4}} \times \left(26.2 + 0.065 tanh\left(\frac{105}{70}\right)(144 - 115)\right)$$

$$F_{bfRd,1} = 10184\,\text{N}$$

$$F_{bfRd,2} = 0.95 \times 20 \times 6 \times 150^{\frac{1}{4}} \times \left(26.2 + 0.065 tanh\left(\frac{150}{70}\right)(144 - 115)\right)$$

$$F_{bfRd,2} = 11\,184\ \text{N}$$

$$F_{bfRd} = 2F_{bfRd,1} + 2F_{bfRd,2} = 2 \times 10184 + 2 \times 11184 = 42\,736\ \text{N} = 42.7\ \text{kN}$$

In this case a_l is calculated using

$$d_{equiv} = \frac{1213 \times 170\,000 \times 200 + (1143 + 1079) \times 200\,000 \times 4020}{170\,000 \times 200 + 200\,000 \times 8040} = 1113\ \text{mm}$$

Since the anchorage region is within the cracked zone of the beam, an anchorage length of $2s_{cr,\,min}$ will be provided to ensure that the bond resistance can be generated between cracks.

As indicated in Figure 13.13, the resulting overall CFRP strengthening length is
$(12.96 - 3.04) + 2 \times (2 \times 0.144) = 10.5$ m (at $\alpha_{fE} = 2.75$ m distance from centre of support)

Figure 13.13 Force in CFRP and end anchorage resistance – NSM strengthening separation

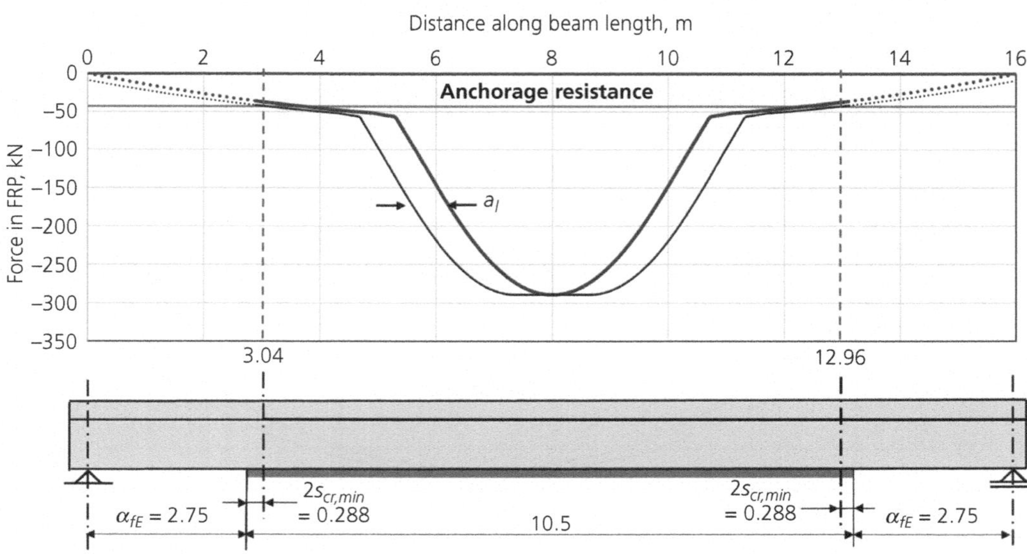

Step 4.2 – End cover separation check (*BS EN 1992-1-1:2023*, J.11.1.3 (2))
The end cover separation shall be checked to *BS EN 1992-1-1:2023*, J.11.1.2.4.

Calculate the maximum shear force at the end of CFRP reinforcement at distance $\alpha_{fE} = 2.75$ m from centre of support

$$V_{Ed} = 535.5 \text{ kN}$$

$$V_{Rd,cfE} = \left(0.11 + 2.2\frac{(100\rho_l)^{0.15}}{\alpha_{fE}^{0.36}}\right)(100\rho_l f_{ck})^{\frac{1}{3}} b_w d_{st}$$

$$\rho_l = \frac{A_{sl}}{b_w \times d_{st}} = \frac{8040}{510 \times 1111} = 0.0142 \text{ (\textit{BS EN 1992-1-1:2023}, Formula 8.28)}$$

$$V_{Rd,cfE} = \left(0.11 + 2.2 \times \frac{(100 \times 0.0142)^{0.15}}{2750^{0.36}}\right) \times (100 \times 0.0142 \times 30)^{\frac{1}{3}} \times 510 \times 1111$$

$$V_{Rd,cfE} = 482\,703 \text{ N} = 483 \text{ kN}$$

$$V_{Ed} = 535.5 \text{ kN} > V_{Rd,cfE} = 483 \text{ kN}$$

Therefore, shear strengthening at the end of the longitudinal strengthening shall be provided to resist the force calculated in accordance with Equation 3.16

$$F_{st} = F_{bfRd} tan\theta$$

Angle θ is taken as per Example Flexural Design B-1.

$$F_{st} = 42.7 \times tan 39.8° = 35.6 \text{ kN}$$

The force is similar to the one calculated in Example Flexural Design B-2 and, therefore, similar end stirrup design will be adopted.

Using 4 layers of continuous CF sheets, the required area of end stirrup will be

$$A_f = \frac{35.6}{250} \times 10^3 = 142.2 \text{ mm}^2$$

The required width per leg will be

$$b_f = \frac{142.2}{2 \times 0.543} = 131 \text{ mm}$$

2 legs of 4 layers of continuous CF sheets with 131 mm width will be adequate, as indicated in Figure 13.14.

Figure 13.14 End shear stirrup to prevent end cover separation – NSM strengthening – all dimensions are in mm

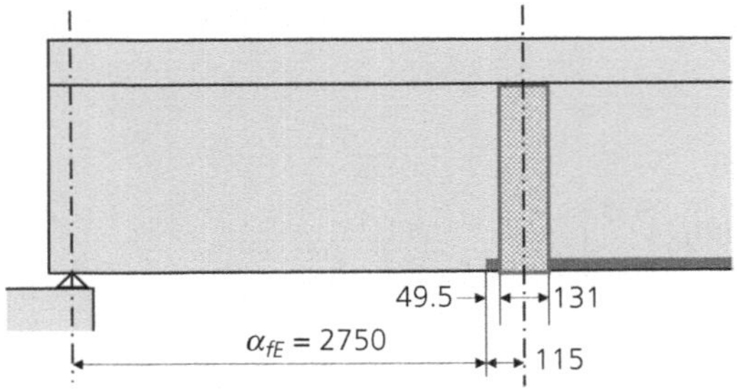

Step 4.3 – Shear induced separation check (*BS EN 1992-1-1:2023*, J.11.1.3 (3))

$$V_{Ed} = 535.5 \text{ kN}$$

Calculate the design shear stress at the end of CFRP reinforcement in accordance with *BS EN 1992-1-1:2023*, Formula 8.18.

$$\tau_{Ed} = \frac{V_{Ed}}{b_w z} = \frac{V_{Ed}}{b_w 0.9 d_{st}} = \frac{535.5 \times 10^3}{510 \times 0.9 \times 1111} = 1.05 \text{ MPa}$$

Check if the limit in *BS EN 1992-1-1: 2023*, Formula J.54 is exceeded.

$$\tau_{Ed} = 1.05 \text{ MPa} < 0.33 f_{ck}^{\frac{2}{3}} = 3.2 \text{ MPa – check is ok.}$$

$$\tau_{Rd} = 4.18 \text{ MPa – as per Example Flexural Design B-1}$$

As the check is undertaken outside of the region close to the support, the stress in the steel stirrups may be calculated by equally distributing the applied shear force into each leg of each stirrup that crosses the shear crack at the prescribed angle.

$$\sigma_{swd} = \frac{V_{Ed}}{\left(\frac{A_{sw}}{s}\right) z\left(cot\theta + cot\alpha\right) sin\alpha} = \frac{535.5 \times 10^3}{\left(\frac{402}{200}\right) \times 0.9 \times 1111 \times \left(1.2 + 0\right) \times 1} = 222 \text{ MPa}$$

$$\frac{\tau_{Ed}\sigma_{swd}}{\tau_{Rd}} = \frac{1.05 \times 222}{4.18} = 55.9 \text{ MPa} < 75 \text{ MPa}$$

Therefore, additional CFRP stirrups are not required.

The NSM strengthening provides a more efficient solution in comparison with Example Flexural Design B-2, while using the same material system.

Kaloyana Kostova and Craig Giaccio
ISBN 978-1-83549-935-1
https://doi.org/10.1108/978-1-83549-934-420251015

Annex C
Crack width calculation to fib Bulletin No. 90

fib Bulletin No. 90 (fib, 2019) provides guidance on crack width calculation, which follows an iterative approach described in Figure 14.1.

Figure 14.1 Flow chart summarising the crack width calculation provided in fib Bulletin No. 90 (fib, 2019)

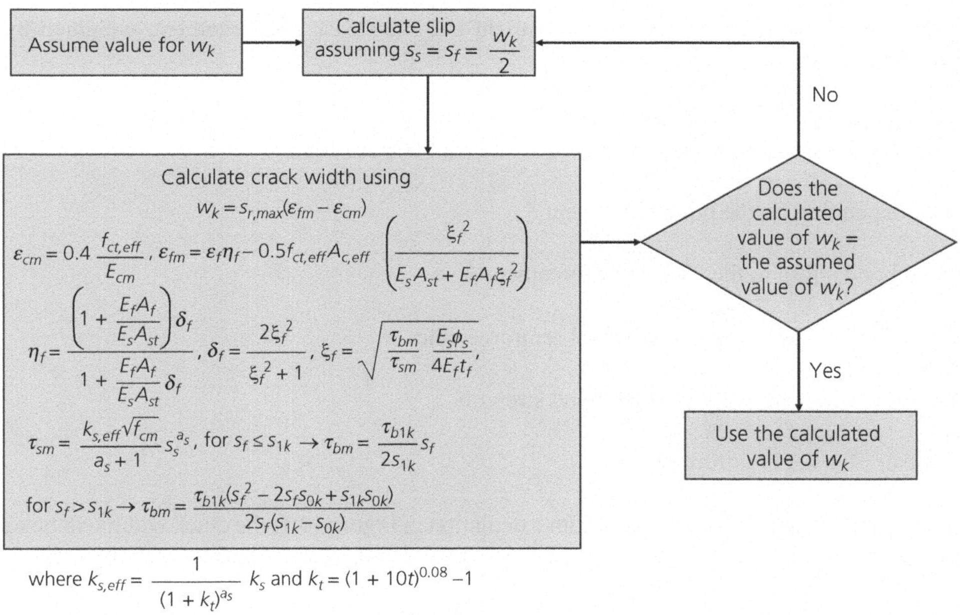

Some symbols used in fib Bulletin No. 90 differ from those defined in *BS EN 1992-1-1:2023* (BSI, 2023b) and are defined as follows.

a_s Bond factor for internal steel reinforcement

k_s Bond factor for internal steel reinforcement

$k_{s,eff}$ Factor dependent on the duration of the load and the bond conditions of the internal steel reinforcement

k_t Factor dependent on the duration of the load

s_f Slip in the CFRP reinforcement

s_s Slip in the steel reinforcement

$s_{r,max}$ Maximum crack spacing

s_{0k} Maximum slip in the CFRP reinforcement

s_{1k} Slip in the CFRP reinforcement at maximum bond stress

t Residual service life span in hours

w_k Upper characteristic value of the crack width of a member strengthened with CFRP

ε_{fm} Mean strain in the CFRP reinforcement between cracks

η_f Strain ratio of the internal steel reinforcement and the CFRP reinforcement, weighted by the respective axial stiffnesses

ξ_f Bond coefficient describing the difference in the bond behaviour of steel and CFRR reinforcement derived from the boundary conditions of the single crack state

δ_f Factor depending on the bond coefficient ξ_f

τ_{bm} Mean bond stress in the CFRP reinforcement

τ_{sm} Mean bond stress in the internal steel reinforcement

τ_{b1k} Characteristic value of the CFRP bond strength

$\varnothing_s$ Diameter of a steel reinforcing bar

Using the iterative approach for crack width calculation in Figure 14.1, the crack width will be calculated for example EBR serviceability 1 (see Chapter 7) assuming a residual service life of 60 years.

First assume crack width w_k in order to calculate slip of steel and CFRP reinforcement $s_s = s_f = \frac{w_k}{2}$

Then calculate $w_k = s_{r,max}\left(\varepsilon_{fm} - \varepsilon_{cm}\right)$ and repeat until assumed and calculated values match.

Assume: $w_k = 0.3$ mm

$$s_s = s_f = 0.15 \text{ mm}$$

$$f_{ct,eff} = f_{fctm} = 3.2 \text{ MPa}$$

$$A_{c,eff} = bh_{c,eff}$$

Where $h_{c,eff} = min\left[2.5(h - d_{st}); \frac{(d_{st} - x_{post})}{3}; \frac{h}{2}\right]$ is taken from BS EN 1992-1-1:2004 (BSI, 2004a).

$$h_{c,eff} = min\left[2.5 \times (250 - 192); \frac{(192 - 62.2)}{3}; \frac{250}{2}\right] = (145; 43.3; 125) = 43.3 \text{ mm}$$

$$A_{c,eff} = 1000 \times 43.3 = 43300 \text{ mm}^2$$

$$t = 60 \times 365 \times 24 = 525600 \text{ h}$$

Bond factors (for ribbed bars and medium bond conditions)

$k_s = 1.0$ and $a_s = 0.25$

Therefore

$$k_t = (1 + 10t)^{0.08} - 1 = (1 + 10 \times 525600)^{0.08} - 1 = 2.449$$

$$k_{s,eff} = \frac{1}{(1 + k_t)^{a_s}} k_s = \frac{1}{(1 + 2.449)^{0.25}} \times 1.0 = 0.734$$

$$\tau_{sm} = \frac{k_{s,eff}\sqrt{f_{cm}}}{a_s + 1} s_s^{\,a_s} = \frac{0.734 \times \sqrt{43}}{0.25 + 1} \times 0.15^{0.25} = 2.40 \text{ MPa}$$

$$s_{0k} = 0.2 \text{ mm}$$

$$\tau_{b1k} = \tau_{f1k} = 0.37\sqrt{f_{cm} f_{ctm}} = 0.37 \times \sqrt{43 \times 3.2} = 4.34 \text{ MPa}$$

$$s_{1k} = 2.5 \frac{50}{E_{cm}} \tau_{b1k} = 2.5 \times \frac{50}{33282} \times 4.34 = 0.016 \text{ mm}$$

$s_f = 0.15 > s_{1k}$, therefore

$$\tau_{bm} = \frac{\tau_{b1k}\left(s_f^{\,2} - 2s_f s_{0k} + s_{1k} s_{0k}\right)}{2s_f\left(s_{1k} - s_{0k}\right)} = \frac{4.34 \times \left(0.15^2 - 2 \times 0.15 \times 0.2 + 0.016 \times 0.2\right)}{2 \times 0.15 \times (0.016 - 0.2)} = 2.70 \text{ MPa}$$

$$\xi_f = \sqrt{\frac{\tau_{bm}}{\tau_{sm}} \frac{E_s \varnothing_s}{4E_f t_f}} = \sqrt{\frac{2.70}{2.40} \times \frac{200000 \times 16}{4 \times 170000 \times 1.2}} = 2.1$$

$$\delta_f = \frac{2\xi_f^{\,2}}{\xi_f^{\,2} + 1} = \frac{2 \times 2.1^2}{2.1^2 + 1} = 1.63$$

$$\eta_f = \frac{\left(1 + \frac{E_f A_f}{E_s A_{st}}\right)\delta_f}{1 + \frac{E_f A_f}{E_s A_{st}}\delta_f} = \frac{\left(1 + \frac{170\,000 \times 384}{200\,000 \times 2010}\right) \times 1.63}{1 + \frac{170\,000 \times 384}{200\,000 \times 2010} \times 1.63} = 1.5$$

$$\varepsilon_{cm} = 0.4\frac{f_{ct,eff}}{E_{cm}} = 0.4 \times \frac{3.2}{33\,282} = 3.846 \times 10^{-5}$$

$$\varepsilon_{fm} = \varepsilon_f \eta_f - 0.5 f_{ct,eff} A_{c,eff} \left(\frac{\xi_f^{\ 2}}{E_{st} A_s + E_f A_f \xi_f^{\ 2}}\right)$$

$$\varepsilon_f = \frac{\sigma_f}{E_f} = \frac{190.3}{170\,000} = 0.001$$

$$\varepsilon_{fm} = 0.001 \times 1.5 - 0.5 \times 3.2 \times 43\,300 \times \left(\frac{2.1^2}{200\,000 \times 2010 + 170\,000 \times 384 \times 2.1^2}\right) = 1.235 \times 10^{-3}$$

Therefore, $\varepsilon_{fm} - \varepsilon_{cm} = 1.196 \times 10^{-3}$

$$s_{r,max} = \frac{f_{ct,eff} A_{c,eff}}{2\tau_{bm}} \frac{4E_f t_f \xi_f^{\ 2}}{E_s A_s + E_f A_f \xi_f^{\ 2}}$$

$$= \frac{3.2 \times 43\,300}{2 \times 2.70} \times \frac{4 \times 170\,000 \times 1.2 \times 2.1^2}{200\,000 \times 2010 + 170\,000 \times 384 \times 2.1^2} = 134 \text{ mm}$$

$$w_k = 13 \times 1.196 \times 10^{-3} = 0.16 \text{ mm} < 0.3 \text{ mm}$$

By decreasing the assumed value, it is found that the crack width is

$$w_k = 0.157 \text{ mm}$$

$$\varepsilon_{fm} - \varepsilon_{cm} = 1.179 \times 10^{-3}$$

$$s_{r,max} = 133 \text{ mm}$$

Kaloyana Kostova and Craig Giaccio
ISBN 978-1-83549-935-1
https://doi.org/10.1108/978-1-83549-934-420251016

Annex D
Determination of compression force and lever arm at a section using *BS EN 1992-1-1:2023*, Formula 8.4

Assume a rectangular cross section predominantly subject to flexure of width b as shown in Figure 15.1.

Figure 15.1 Arbitrary rectangular cross section

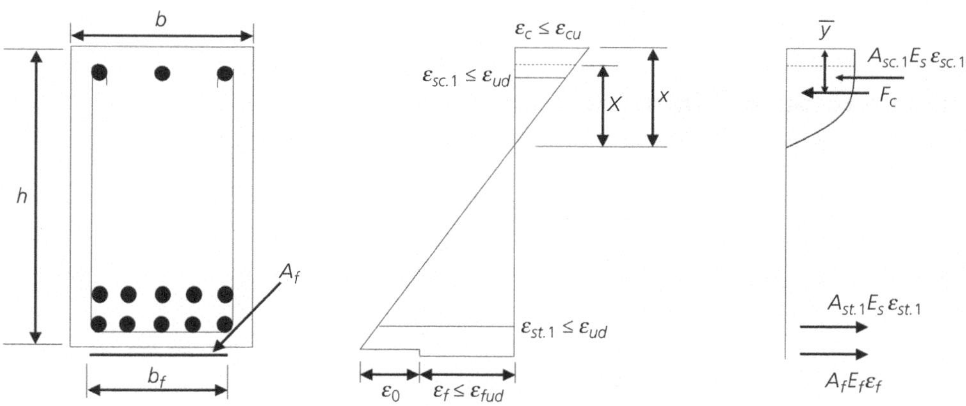

Denotes location in cross section at which concrete compressive strain $\varepsilon_c = \varepsilon_{c2}$

At ultimate limit state (ULS), *BS EN 1992-1-1:2023* (2023b), Formula 8.4 holds as follows.

$$\sigma_c = \begin{cases} f_{cd}\left[1-\left(1-\dfrac{\varepsilon_c}{\varepsilon_{c2}}\right)^2\right], & \text{where } \varepsilon_c \le \varepsilon_{c2} \\ f_{cd}, & \text{where } \varepsilon_{c2} \le \varepsilon_c \le \varepsilon_{cu} \end{cases}$$

Where

$\varepsilon_{c2} = 0.002$

$\varepsilon_{cu} = 0.0035$

It will be demonstrated that the resulting compression force from this constitutive model for a section predominantly subject to flexure can be expressed with Equation 15.1 and the distance of the force resultant to the outermost compression fibre, $\overline{y}$, can be determined using Equation 15.3.

$$F_c = k_1 f_{cd}\, bx \tag{15.1}$$

Where

$$k_1 = \begin{cases} 1000\varepsilon_c\left[0.5 - \dfrac{250\varepsilon_c}{3}\right], \ \varepsilon_c < 0.002 \\ 1 - \dfrac{1}{1500\varepsilon_c}, \qquad 0.002 \le \varepsilon_c \le 0.0035 \end{cases} \tag{15.2}$$

$$\frac{\overline{y}}{x} = k_2 = \begin{cases} \dfrac{2 - 250\varepsilon_c}{6 - 1000\varepsilon_c}, \ \varepsilon_c < 0.002 \\ \dfrac{1000\varepsilon_c\left(3000\varepsilon_c - 4\right) + 2}{2000\varepsilon_c\left(3000\varepsilon_c - 2\right)}, \ 0.002 \le \varepsilon_c \le 0.0035 \end{cases} \tag{15.3}$$

Approach to determining solution

To determine the force in the compression region F_c, the following applies.

$$F_c = \int_0^x b\sigma_c\left(y\right)dy \tag{15.4}$$

Where

$$\sigma_c\left(y\right) = \begin{cases} f_{cd}\left[1 - \left(1 - \dfrac{\varepsilon_c\left(y\right)}{\varepsilon_{c2}}\right)^2\right], \text{where } \varepsilon_c \le \varepsilon_{c2} \\ f_{cd}, \qquad \text{where } \varepsilon_{c2} \le \varepsilon_c \le \varepsilon_{cu} \end{cases} \tag{15.5}$$

The distance of the force from the outermost compression fibre ($\overline{y}$ in Figure 15.1) can be calculated using

$$\overline{y} = k_2 x = \frac{\int_0^x by\sigma_c\left(y\right)dy}{\int_0^x b\sigma_c\left(y\right)dy} \tag{15.6}$$

Where k_2 is defined in Equation 15.3.

This is to be solved in two cases.

Case 1: $\varepsilon_c \le \varepsilon_{c2}$

Case 2: $\varepsilon_{c2} \le \varepsilon_c \le \varepsilon_{cu}$

Case 1: $\varepsilon_c \leq \varepsilon_{c2}$

Derivation of compression force

From Equation 15.4 and Equation 15.5 we obtain the following for integration.

$$F_c = \int_0^x bf_{cd}\left[1 - \left(1 - \frac{\varepsilon(y)}{\varepsilon_{c2}}\right)^2\right]dy \tag{15.7}$$

The compressive strain, ε, varies linearly with the depth of the compression zone from $\varepsilon = 0$ at the neutral axis to the maximum value of strain at the outermost compressive fibre of the section such that $\varepsilon \leq \varepsilon_c$. In this case, $\varepsilon_c \leq \varepsilon_{c2}$. This can be written as

$$\varepsilon(y) = \frac{y}{x}\varepsilon_c \tag{15.8}$$

Substituting Equation 15.8 into Equation 15.7 and removing constants from integration yields

$$F_c = bf_{cd}\int_0^x \left[1 - \left(1 - \frac{y\varepsilon_c}{x\varepsilon_{c2}}\right)^2\right]dy \tag{15.9}$$

Expanding and simplifying the integrand, the following is obtained

$$F_c = bf_{cd}\left[\frac{2\varepsilon_c}{x\varepsilon_{c2}}\int_0^x ydy - \left(\frac{\varepsilon_c}{x\varepsilon_{c2}}\right)^2 \int_0^x y^2 dy\right] \tag{15.10}$$

Solving these relatively straightforward integrals and simplifying yields

$$F_c = bf_{cd}\varepsilon_c \mathrm{x}\left[\frac{1}{\varepsilon_{c2}} - \frac{\varepsilon_c}{3\varepsilon_{c2}{}^2}\right] \tag{15.11}$$

Given that $\varepsilon_{c2} = 0.002$

$$F_c = 1000bf_{cd}\varepsilon_c x\left[0.5 - \frac{250\varepsilon_c}{3}\right] \tag{15.12}$$

Comparing Equation 15.12 to Equation 15.1, k_1 is found to be

$$k_1 = 1000\varepsilon_c\left[0.5 - \frac{250\varepsilon_c}{3}\right] \text{for } \varepsilon_c \leq \varepsilon_{c2} \tag{15.13}$$

Derivation of the distance of the force from the outermost compression fibre

To find the distance of the line of action of the force from the outermost compression fibre, we solve Equation 15.6 as below (noting that the common term of width cancels from the numerator and denominator).

$$\bar{y} = k_2 x = \frac{\int_0^x y\sigma_c(y)\,dy}{\int_0^x \sigma_c(y)\,dy}$$

The denominator is given by Equation 15.12, so therefore we only need to find the numerator.

The numerator can be written as

$$\int_0^x y\sigma_c(y)\,dy = \int_0^x bf_{cd}\left[y - y\left(1 - \frac{\varepsilon(y)}{\varepsilon_{c2}}\right)^2\right]dy$$

Giving

$$\int_0^x y\sigma_c(y)\,dy = bf_{cd}\left[\int_0^x 2y\frac{\varepsilon(y)}{\varepsilon_{c2}} - y\left(\frac{\varepsilon(y)}{\varepsilon_{c2}}\right)^2\right]dy \tag{15.14}$$

Taking again that $\varepsilon(y) = \frac{y}{x}\varepsilon_c$ and integrating this numerator using a similar method as above and substituting this along with Equation 15.12 into Equation 15.6 yields

$$\bar{y}_{NA} = x\left(\frac{4 - 750\varepsilon_c}{6 - 1\,000\varepsilon_c}\right)$$

Noting that the formula for $\varepsilon(y)$ yields the position of the force relative to the neutral axis (hence the terminology $\bar{y}_{NA}$), the following needs to be done to obtain the position of the force relative to the outermost compression fibre

$$\bar{y} = x - \bar{y}_{NA}$$

Which yields

$$\bar{y} = x\left(\frac{2 - 250\varepsilon_c}{6 - 1000\varepsilon_c}\right) \tag{15.15}$$

Comparing Equation 15.15 to Equation 15.3, k_2 is found to be

$$k_2 = \frac{2 - 250\varepsilon_c}{6 - 1000\varepsilon_c} \text{ for } \varepsilon_c \le \varepsilon_{c2} \tag{15.16}$$

Case 2: $\varepsilon_{c2} \leq \varepsilon_c \leq \varepsilon_{cu}$

Figure 15.2 shows the compression region under consideration, excluding compression reinforcement.

Derivation of compression force

To find the compression force, we again solve Equation 15.4.

As shown in Figure 15.2, in this case, there are two regions as follows.

Region 1 where $\varepsilon_{c2} \leq \varepsilon_c \leq \varepsilon_{cu}$, and the design compressive stress is constant and equal to f_{cd}, yielding Equation 15.17.

$$F_{c1} = b\int_{X}^{x} f_{cd}\,dy \tag{15.17}$$

Region 2 where $\varepsilon_c \leq \varepsilon_{c2}$, yielding Equation 15.18.

$$F_{c2} = b\int_{0}^{X} f_{cd}\left[1-\left(1-\frac{\varepsilon(y)}{\varepsilon_{c2}}\right)^{2}\right]dy \tag{15.18}$$

Where

X is the distance from the neutral axis to the location in the cross section where the compressive strain $\varepsilon_X = \varepsilon_{c2}$

The total compression force C can be found using Equation 15.19.

$$F_c = F_{c1} + F_{c2} \tag{15.19}$$

This can again be solved in two parts.

Figure 15.2 Stresses and strains in compression region for case 2

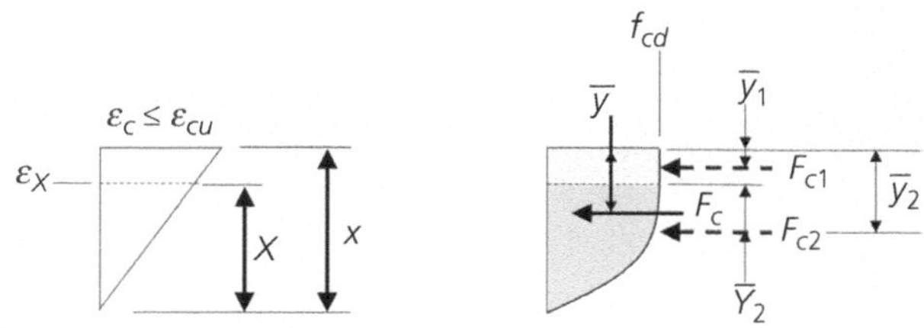

Denotes location in cross section at which concrete compressive strain $\varepsilon_c = \varepsilon_{c2}$

Denotes Region 1

Denotes Region 2

Region 1

As given in Equation 15.17, the stress is constant through this part of the section.

$$F_{c1} = b\int_{X}^{x} f_{cd}dy$$

Therefore

$$F_{c1} = bf_{cd}\left[y\right]_{X}^{x} = bf_{cd}\left(x - X\right) \tag{15.20}$$

$$X = \frac{\varepsilon_{c2}}{\varepsilon_c}x \tag{15.21}$$

Substituting this back into Equation 15.20, simplifying, and substituting in the known value of $\varepsilon_{c2} = 0.002$ yields

$$F_{c1} = bf_{cd}x\left(1 - \frac{1}{500\varepsilon_c}\right) \tag{15.22}$$

Comparing Equation 15.22 to Equation 15.1, k_1 is found to be

$$k_1 = \left(1 - \frac{1}{500\varepsilon_c}\right) \text{ for } \varepsilon_{c2} \leq \varepsilon_c \leq \varepsilon_{cu} \tag{15.23}$$

Region 2

Region 2 involves solving Equation 15.18, which is similar to solving Equation 15.7.

With reference to Figure 15.2, the solution to Equation 15.7 shown in Equation 15.11 can be adapted as follows.

$$F_{c2} = bf_{cd}\varepsilon_X X\left(\frac{1}{\varepsilon_{c2}} - \frac{\varepsilon_X}{3\varepsilon_{c2}^{\ 2}}\right)$$

In this case, $\varepsilon_X = \varepsilon_{c2}$, therefore

$$F_{c2} = bf_{cd}X\left(1 - \frac{1}{3}\right)$$

That is,

$$F_{c2} = \frac{2}{3}bf_{cd}X$$

$$F_{c2} = \frac{2}{3}bf_{cd}X$$

Given that $X = \frac{\varepsilon_{c2}}{\varepsilon_c} x$, and $\varepsilon_{c2} = 0.002$

$$F_{c2} = \frac{2xbf_{cd}}{1500\varepsilon_c} \tag{15.24}$$

Substituting Equation 15.22 and Equation 15.24, Equation 15.19 yields a total compression force on the cross section of

$$F_c = bf_{cd}x\left(1 - \frac{1}{500\varepsilon_c}\right) + \frac{2xbf_{cd}}{1500\varepsilon_c}$$

Simplifying, this yields

$$F_c = bf_{cd}x\left[1 - \frac{1}{1500\varepsilon_c}\right] \tag{15.25}$$

Derivation of the distance of the force from the outermost compression fibre

To find the distance of the line of action of the force from the outermost compression fibre, we again solve Equation 15.6 as below (noting that the common term of width cancels from the numerator and denominator).

The two regions used for the compression force will again be used for the determination of this distance, with the distance of the line of action of the force from the outermost compression fibre for the full compression region determined as

$$\bar{y} = \frac{F_{c1}\bar{y}_1 + F_{c2}\bar{y}_2}{F_{c1} + F_{c2}} \tag{15.26}$$

$$F_{c1} = bf_{cd}x\left(1 - \frac{1}{500\varepsilon_c}\right)$$

$$F_{c2} = \frac{2xbf_{cd}}{1\,500\varepsilon_c}$$

$$F_c = F_{c1} + F_{c2} = bf_{cd}x\left[1 - \frac{1}{1500\varepsilon_c}\right]$$

Region 1

With reference to Figure 15.2, the determination of the distance of the line of action of the force from the outermost compression fibre in this region is simply

$$\bar{y}_1 = \frac{x - X}{2}$$

Given that $X = \frac{\varepsilon_{c2}}{\varepsilon_c}x$, then

$$\bar{y}_1 = \frac{x - \frac{\varepsilon_{c2}}{\varepsilon_c}x}{2}$$

Simplifying, and noting that $\varepsilon_{c2} = 0.002 \left(= \frac{1}{500}\right)$ yields

$$\bar{y}_1 = x\left(\frac{1}{2} - \frac{1}{1000\varepsilon_c}\right) \qquad (15.27)$$

Region 2

In region 2, Equation 15.6 is modified as follows.

$$\overline{y_2} = \frac{yF_c}{F_c} = \frac{\int_0^X yb\sigma_c(y)dy}{\int_0^X b\sigma_c(y)dy} \qquad (15.28)$$

Which is identical to solving Equation 15.14. Replacing ε_c x with ε_X and X respectively in the solution to this equation (Equation 15.15) yields the lever arm relative to the top of region 2 (refer to Figure 15.2)

$$\bar{Y}_2 = X\left(\frac{2 - 250\varepsilon_X}{6 - 1\,000\varepsilon_X}\right) \qquad (15.29)$$

Noting that $\varepsilon_X = \varepsilon_{c2}$, and $\varepsilon_{c2} = 0.002 = \left(\frac{1}{500}\right)$ we obtain

$$\bar{Y}_2 = \frac{3}{8}X \qquad (15.30)$$

$\bar{Y}_2$ is measured from the point at which the strain reaches ε_X. We require $\bar{y}_2$ to solve Equation 15.26. This can be found as follows.

$$\bar{y}_2 = (x - X) + \bar{Y}_2$$

Or

$$\bar{y}_2 = x - \frac{5}{8}X$$

Given that $X = \frac{\varepsilon_{c2}}{\varepsilon_c}x$, $\varepsilon_X = \varepsilon_{c2}$, and $\varepsilon_{c2} = 0.002 = \left(\frac{1}{500}\right)$ we obtain

$$\bar{y}_2 = x - \frac{5}{8}\frac{1}{500\varepsilon_c}x$$

$$\bar{y}_2 = x\left(1 - \frac{1}{800\varepsilon_c}\right) \tag{15.31}$$

To find the resulting distance of the line of action of the compression force for the entire compression region, we recall that

$$\bar{y} = \frac{F_{c1}\bar{y}_1 + F_{c2}\bar{y}_2}{F_{c1} + F_{c2}} \tag{15.32}$$

Substituting Equation 15.22, Equation 15.24, Equation 15.27 and Equation 15.31 into Equation 15.32 yields

$$\bar{y} = x\left(\frac{\frac{1}{2}\left(1 - \frac{1}{500\varepsilon_c}\right)^2 + \frac{1}{750\varepsilon_c}\left(1 - \frac{1}{800\varepsilon_c}\right)}{\left(1 - \frac{1}{1500\varepsilon_c}\right)}\right)$$

To simplify, we expand as follows

$$\bar{y} = x\left(\frac{\frac{1}{2}\left(\frac{500\varepsilon_c - 1}{500\varepsilon_c}\right)^2 + \frac{1}{750\varepsilon_c}\left(\frac{800\varepsilon_c - 1}{800\varepsilon_c}\right)}{\left(\frac{1500\varepsilon_c - 1}{1500\varepsilon_c}\right)}\right)$$

Simplifying this yields

$$\bar{y} = 1500x\left[\frac{\frac{(500\varepsilon_c - 1)^2}{500\,000\varepsilon_c} + \frac{(800\varepsilon_c - 1)}{600\,000\varepsilon_c}}{(1500\varepsilon_c - 1)}\right]$$

Which gives

$$\bar{y} = x\left[\frac{\frac{3(500\varepsilon_c - 1)^2}{1000\varepsilon_c} + \frac{(800\varepsilon_c - 1)}{400\varepsilon_c}}{(1500\varepsilon_c - 1)}\right]$$

Multiplying the second term on the numerator by $\frac{4}{4}$ and taking grouping terms, the following is obtained

$$\frac{\bar{y}}{x} = \frac{3(500\varepsilon_c - 1)^2 + 2.5(800\varepsilon_c - 1)}{1000\varepsilon_c(1500\varepsilon_c - 1)}$$

Multiplying the entire expression by $\frac{2.5}{2.5}$, the following is found

$$\frac{\bar{y}}{x} = \frac{12(500\varepsilon_c - 1)^2 + 10(800\varepsilon_c - 1)}{2000\varepsilon_c(3000\varepsilon_c - 2)}$$

Expanding the numeration yields

$$\frac{\bar{y}}{x} = \frac{3 \times 10^6 \varepsilon_c{}^2 - 4000\varepsilon_c + 2}{2000\varepsilon_c(3000\varepsilon_c - 2)}$$

Further simplifying the numerator, and noting that $k_2 = \frac{\bar{y}}{x}$ the following is obtained

$$k_2 = \frac{\bar{y}}{x} = \frac{1000\varepsilon_c(3000\varepsilon_c - 4) + 2}{2000\varepsilon_c(3000\varepsilon_c - 2)} \quad (15.33)$$

emerald PUBLISHING ice Publishing

Kaloyana Kostova and Craig Giaccio
ISBN 978-1-83549-935-1
https://doi.org/10.1108/978-1-83549-934-420251017

Annex E
Determination of compression force and lever arm at a section using *BS EN 1992-1-1:2023*, Formula 5.6

The general stress-strain curve for concrete is given in *BS EN 1992-1-1:2023* (BSI, 2023b), 5.1.6 as Equation 16.1.

$$\sigma_c = f_{cm}\frac{k\eta - \eta^2}{1+(k-2)\eta} \tag{16.1}$$

Where

$$k = 1.05E_{cm}\frac{\varepsilon_{c1}}{f_{cm}} \tag{16.2}$$

$$\varepsilon_{c1} = 0.7f_{cm}^{\frac{1}{3}} \tag{16.3}$$

$$\eta = \frac{\varepsilon_c}{\varepsilon_{c1}} \tag{16.4}$$

ε_c is the maximum compression strain in the cross section

$$E_{cm} = 9500f_{cm}^{\frac{1}{3}} \tag{16.5}$$

Note that Equation 16.5 uses the recommended value of the adjustment factor for the aggregate type from *BS EN 1992-1-1:2023* of $k_E = 9500$.

For a section predominantly subject to flexure, the resulting compression force F_c, and the distance of the neutral axis from the outermost compression fibre, $\overline{y}$, are as provided below. Nomenclature is as outlined in Figure 15.1.

$$F_c = bxf_{cm}\left(kk_2\varepsilon_c I_1 - k_2^2\varepsilon_c^2 I_2\right) \tag{16.6}$$

$$\overline{y} = x\frac{\left(kk_2\varepsilon_c(I_1 - I_2) - k_2^2\varepsilon_c^2(I_2 - I_3)\right)}{kk_2\varepsilon_c I_1 - k_2^2\varepsilon_c^2 I_2} \tag{16.7}$$

Where

$$I_1 = \frac{1}{A^2}\left[A - \text{In}\left(A+1\right)\right] \tag{16.8}$$

$$I_2 = \frac{1}{A^3}\left[\frac{A^2}{2} - A + \text{In}\left(A+1\right)\right] \tag{16.9}$$

$$I_3 = \frac{1}{A^4}\left[\frac{A^3}{3} - \frac{A^2}{2} + A - \text{In}\left(A+1\right)\right] \tag{16.10}$$

Where

$$k_2 = \frac{1}{\varepsilon_{c1}}$$

$$A = \left(k-2\right)k_2\varepsilon_c$$

Note this method can be applied to ultimate limit state (ULS) by considering substitutions in Section 7.4.3.3 of *BS EN 1992-1-1:2023*, which requires that

- f_{cm} is replaced by f_{cd}
- E_{cm} is replaced by E_{cd}

Where

$$E_{cd} = \frac{E_{cd}}{\gamma_{cE}} \tag{16.11}$$

γ_{cE} is the partial factor for the modulus of elasticity taken from Table 4.1 in *BS EN 1992-1-1:2023* as $\gamma_{cE} = 1.3$.

Derivation of compression force

The compression force of a rectangular section can be found by integrating Equation 16.1 throughout the depth of the compression region as follows.

$$F_c = b\int_{y=0}^{y=x} \sigma\left(y\right)dy$$

In the stress-strain relationship, it is the term η that varies with depth, therefore Equation 16.1 can be rewritten as

$$\sigma_c\left(y\right) = f_{cm}\frac{k\eta\left(y\right) - \eta\left(y\right)^2}{1+\left(k-2\right)\eta\left(y\right)} \tag{16.12}$$

Where

$$\eta(y) = \frac{1}{\varepsilon_{c1}} \varepsilon(y)$$

In this case, take the strain from the outermost compressive fibre such that

$$\varepsilon(y) = \varepsilon_c \frac{x - y}{x}$$

Let

$$k_2 = \frac{1}{\varepsilon_{c1}}$$

$$A = (k - 2) k_2 \varepsilon_c$$

Substituting this into Equation 16.12 yields

$$\sigma_c(y) = f_{cm} \frac{k k_2 \varepsilon_c \frac{x - y}{x} - k_2^2 \varepsilon_c^2 \left(\frac{x - y}{x}\right)^2}{1 + A\left(\frac{x - y}{x}\right)}$$

This is difficult to integrate, so the substitution $u = \frac{x - y}{x}$ is made

This also yields $dy = -xdu$

Making this additional substitution yields

$$\sigma_c(y) = f_{cm} \frac{k k_2 \varepsilon_c u - k_2^2 \varepsilon_c^2 u^2}{1 + Au} \tag{16.13}$$

Now

$$F_c = b \int_{y=0}^{y=x} \sigma(y) dy$$

Or

$$F_c = b f_{cm} \int_{y=0}^{y=x} \frac{k k_2 \varepsilon_c u - k_2^2 \varepsilon_c^2 u^2}{1 + Au} (-xdu)$$

At y=0, u=1, and at y=x, u=0, also, reversing the limits of integration simply results in multiplying the expression by -1, therefore

$$F_c = x b f_{cm} \int_{u=0}^{u=1} \frac{k k_2 \varepsilon_c u - k_2^2 \varepsilon_c^2 u^2}{1 + Au} du$$

Separate this into two integrals

$$F_c = xbf_{cm}\left[kk_2\varepsilon_c\int_{u=0}^{u=1}\frac{u}{1+Au}du - k_2{}^2\varepsilon_c{}^2\int_{u=0}^{u=1}\frac{u^2}{1+Au}du\right]$$

Simplifying

$$F_c = xbf_{cm}\left[kk_2\varepsilon_c I_1 - k_2{}^2\varepsilon_c{}^2 I_2\right] \tag{16.14}$$

Where

$$I_1 = \int_{u=0}^{u=1}\frac{u}{1+Au}du$$

$$I_2 = \int_{u=0}^{u=1}\frac{u^2}{1+Au}du$$

The solution to the two integrals requires further substitution of $v = 1 + Au$. Performing this substitution and solving gives the following solutions

$$I_1 = \frac{1}{A^2}\left[A - \ln(A+1)\right]$$

$$I_2 = \frac{1}{A^3}\left[\frac{A^2}{2} - A + \ln(A+1)\right]$$

Which produces the solution given in Equation 16.6.

Derivation of lever arm

To obtain the lever arm for a rectangular section, the moment produced by the compression force is divided by the magnitude of the compression force as outlined in Equation 16.15.

$$\bar{y} = \frac{b\int_{y=0}^{y=x} y\sigma(y)dy}{b\int_{y=0}^{y=x}\sigma(y)dy} \tag{16.15}$$

The denominator of this simply corresponds to the compression force summarised in Equation 16.6, and derived above.

Consider the numerator, neglect the width term as it cancels.

Using Equation 16.13, the following is obtained

$$\int_{y=0}^{y=x} y\sigma(y)dy = \int_{u=0}^{u=1} xf_{cm}\frac{ykk_2\varepsilon_c u - yk_2{}^2\varepsilon_c{}^2u^2}{1+Au}du$$

Given $u = \dfrac{x-y}{x}$

$$y = x(1-u)$$

This yields

$$\int_{y=0}^{y=x} y\sigma(y)\,dy = x^2 f_{cm} \int_{u=0}^{u=1} \frac{kk_2\varepsilon_c u(1-u) - k_2^{\,2}\varepsilon_c^{\,2}u^2(1-u)}{1+Au}\,du$$

This can be rewritten as

$$\int_{y=0}^{y=x} y\sigma(y)\,dy = x^2 f_{cm}\Bigg[kk_2\varepsilon_c \int_{u=0}^{u=1} \frac{u}{1+Au}\,du - kk_2\varepsilon_c \int_{u=0}^{u=1} \frac{u^2}{1+Au}\,du$$
$$-k_2^{\,2}\varepsilon_c^{\,2} \int_{u=0}^{u=1} \frac{u^2}{1+Au}\,du + k_2^{\,2}\varepsilon_c^{\,2} \int_{u=0}^{u=1} \frac{u^3}{1+Au}\,du\Bigg]$$

Recalling that

$$I_1 = \int_{u=0}^{u=1} \frac{u}{1+Au}\,du$$

and

$$I_2 = \int_{u=0}^{u=1} \frac{u^2}{1+Au}\,du$$

This can be rewritten as

$$\int_{y=0}^{y=x} y\sigma(y)\,dy = x^2 f_{cm}\left[kk_2\varepsilon_c I_1 - kk_2\varepsilon_c I_2 - k_2^{\,2}\varepsilon_c^{\,2} I_2 + k_2^{\,2}\varepsilon_c^{\,2} I_3\right]$$

Where

$$I_3 = \int_{u=0}^{u=1} \frac{u^3}{1+Au}\,du$$

Integrating I_3 using the substitution $v = 1 + Au$ yields

$$I_3 = \frac{1}{A^4}\left[\frac{A^3}{3} - \frac{A^2}{2} + A - \ln(A+1)\right]$$

And further simplified to

$$\int_{y=0}^{y=x} y\sigma(y)\,dy = x^2 f_{cm}\left[kk_2\varepsilon_c(I_1 - I_2) - k_2^{\,2}\varepsilon_c^{\,2}(I_2 - I_3)\right]$$

Substituting this, and the value for the compression force, into Equation 16.15 yields

$$\bar{y} = \frac{x^2 f_{cm}\left[kk_2\varepsilon_c\left(I_1 - I_2\right) - k_2^2\varepsilon_c^2\left(I_2 - I_3\right)\right]}{xf_{cm}\left[kk_2\varepsilon_c I_1 - k_2^2\varepsilon_c^2 I_2\right]}$$

Which simplifies to

$$\bar{y} = x\frac{kk_2\varepsilon_c\left(I_1 - I_2\right) - k_2^2\varepsilon_c^2\left(I_2 - I_3\right)}{kk_2\varepsilon_c I_1 - k_2^2\varepsilon_c^2 I_2} \tag{16.16}$$

Which proves Equation 16.7.

Kaloyana Kostova and Craig Giaccio
ISBN 978-1-83549-935-1
https://doi.org/10.1108/978-1-83549-934-420251019
Emerald Publishing Limited:

Index

The manufacturer's authorised representative in the EU for product safety is Easy Access System Europe – Mustamäe tee 50, 10621 Tallinn, Estonia, easproject.com, +372 5696 8939, gpsr.requests@easproject.com

Printed and bound by CPI Group (UK) Ltd, Croydon, CR0 4YY
03/09/2026
02209264-0019